Categories, Bundles and Spacetime Topology

Categories, Bundles and Spacetime Topology

C T J Dodson
University of Lancaster

Shiva Publishing Limited

SHIVA PUBLISHING LIMITED
9 Clareville Road, Orpington, Kent BR5 1RU, UK

British Library Cataloguing in Publication Data

Dodson, C T J
Categories, bundles and spacetime topology.
– (Shiva mathematics series; 1).
1. Differential topology
I. Title
514′.7 QA613.6

ISBN 0–906812–01–1

Printed in Great Britain by Devon Print Group, Exeter, Devon

To

Ben Radvan

&

Heinz Corte

For introducing me to research and for their guidance over the ten years 1959-1969, when we worked together on statistical geometry.

Contents

Introduction

This book grew out of assorted lectures, short courses and seminars I have given at Lancaster, Trieste and elsewhere between 1975 and 1979. The choice of material has been largely influenced by my interest in spacetime geometry and by the interests of many, valued friends working in various branches of theoretical physics.

One of the principal mathematical tools of physicists is differential geometry and conversely much benefit has been derived in mathematics from considerations of the formulation of physical theories. Perhaps the most important current topic is bundle theory, because it is now enjoying a central role in both classical relativity and gauge theories. This is reflected in the amount of space devoted to structures on manifolds in Chapter IV, about one third of the book. To save the time of those occupied with applications, Chapter IV begins with a quick reminder of the elementary aspects of manifolds.

Once hooked on bundle theory, a theoretical physicist soon finds that the intricate mixture of algebra and analysis on manifolds spawns vast families of maps going hither and thither among his spaces. Many of these maps turn out to have excellent pedigrees : they are canonical, or natural entities and they give rise to standard, or universal constructions. The right way to view it all is through the theory of categories, and increasingly the language of this theory is appearing in the literature of physics. Chapter II is an introductory course on category theory with many examples, but little in the way of proofs since these are available in the more advanced texts to which, as elsewhere in the book, abundant references are given. The most widely used category in theoretical physics is Top, consisting of topological spaces and continuous maps. For, most experiments are unable to detect discontinuities and so we resort to simple models from topology : small changes have small effects. The category Top is studied in Chapter III. There, after a short recounting of basic ideas, fully detailed proofs are given of the existence and uniqueness of limiting topologies on a given set under various conditions that commonly arise in applications. Thus, the completeness of Top as a category is displayed by construction of products, coproducts, equalizers and coequalizers.

The last part of the book, Chapter V, is a survey of certain global features of spacetime 4-manifolds. There is little here that is actually new except in emphasis, but the material was fragmented rather widely through journals and conference proceedings so it seemed appropriate to bring it together in one place. We consider first the existence of Lorentz structures and how they relate to paracompactness and nowhere-zero vector fields. Next comes orientability and its relation to nowhere-zero differential 4-forms. Orientability is effectively always assured because we can use a covering spacetime, and we see also the relation to time- and space-orientability. The correspondence between spinor structures and parallelizability is outlined and we finish with an account of the geometrical definition and handling of singularities, by bundle completion and modifications thereof.

The book is organized with a simple numbering system for its subsections, facilitating the copious cross references; there is a comprehensive index and a detailed bibliography. Standard abbreviations and conventions are described in Chapter I. Throughout, exercises are disguised as examples and omitted proofs; the omission of a proof is always accompanied by a reference to its location and often also by a hint at the construction.

As is usually the case, I am indebted to more people than I can name here for help at one stage or another with this book. However, I have benefited particularly from discussions with L.W. Flinn, L.J. Sulley, M.J. Slupinski, M. Radivoiovici, C.J.S. Clarke and R.W. Tucker; also, while at the International Centre

For Theoretical Physics, Trieste, I received much encouragement from its Director, Abdus Salam; to all these and to the others with whom I have had discussions at conferences and seminars go my sincere thanks. I wish finally to thank Lin Sulley and David Towers for help in checking the draft, Alexandra Dodson for help with the bibliography and index, and Sylvia Brennan for yet another excellent job of typing.

Kit Dodson
Lancaster, January 1980.

I Preliminaries

Though elementary definitions from topology and differential geometry have been provided for convenience, some familiarity with them will be expected. Also, some experience of sets, vector spaces and basic group theory will be assumed, but the overall algebraic content of our material is slight and the treatment of categories is from first principles. Our main reference text for prerequisite material is the book Tensor Geometry [28] , which gives a leisurely development from vector spaces through topology to curved manifolds and relativistic spacetime.

It is intended that the book be read in the order of its contents. However, someone who needs a quick grasp of bundle theory could go directly to Chapter IV and ignore references to categorical constructions that are not familiar. The spacetime structure in Chapter V relies on material in Chapter IV and the study of the topological category in Chapter III depends on the category theory in Chapter II.

Wherever it seems to help, without undue loss of precision, we shall adopt an intuitive viewpoint in our constructions. So, for a start, we shall assume a common language of elementary set theory. For a splendidly readable summary of set theory see the first part of Devlin [21], and for an appreciation of why set theory is still a lively developing subject read the second part too. It will usually be sufficient for our purposes to assume that our sets belong to some fixed universe of sets; but we also need <u>proper classes</u> (cf. [46]). We shall make use of the usual logical symbols and abbreviations as follows:

$x \in X$	:	"x an element of the set X"
$\forall x$	:	"for all x"
$\exists x$	:	"there exists x"
$\exists! x$	:	"there exists a unique x"
$\nexists x$	:	"there does not exist x"
$\Longrightarrow$	:	"implies" or "then"
$\Longleftrightarrow$	:	"if and only if" or "is equivalent to"
$\emptyset$	:	"the empty set" or "the set with no elements"
$A \subseteq X$	:	"$(\forall x \in A)(x \in X)$" or "A a subset of X"
$A \nsubseteq X$	:	"$(A \subseteq X)$ is false"
$A = X$	:	"$(A \subseteq X)$ and $(X \subseteq A)$"
$A \subset X$	:	"$(A \subseteq X)$ and $(X \neq A)$"
$\{x \in X \mid P(x)\}$	:	"the set of elements of X for which the statement P is true"
$X \backslash A$	:	"the complement of A in X" $\Longleftrightarrow$ "$\{x \in X \mid x \notin A\}$"
$A \cap B$	:	"$\{x \in X \mid (x \in A)$ and $(x \in B)\}$; $A, B \subseteq X$"
$A \cup B$	:	"$\{x \in X \mid (x \in A)$ or $(x \in B)$ or both$\}$; $A, B \subseteq X$"
$f : X \to Y$ $: x \mapsto f(x)$	:	"$(\forall x \in X)(\exists!\ f(x) \in Y)$" $\Longleftrightarrow$ "f is a map from set X to set Y, sending typical x to f(x)"
dom f	:	"the domain of f, X when $f : X \to Y$"
cod f	:	"the codomain of f, Y when $f : X \to Y$"
im f	:	"the image of f, $\{f(x) \mid x \in \text{dom } f\}$"
sub X	:	"the power set of X, $\{A \subseteq X\}$"
$f^{\leftarrow}$	:	"the inverse image map for subsets of cod f; when $f : X \to Y$ then $f^{\leftarrow} : \text{sub } Y \to \text{sub } X : B \mapsto \{x \in X \mid f(x) \in B\}$"
$i : A \hookrightarrow X$	:	"the inclusion map of a subset $A \subset X$"
I_A	:	"the identity map on A"
$f\vert_A$	:	"the restriction of a map f to a subset A of dom f"

$X/\sim$	:	"the quotient of a set X by an equivalence relation $\sim$ "
$[x]_\sim$	:	"the equivalence class containing x "
$\simeq$	:	"is homeomorphic to"
$\equiv$	:	"is equivalent or isomorphic to"
$N, Z, \mathbb{R}, C$	:	"natural, integer, real complex number systems"
$\square$	:	"end (or omission) of proof"

A handy reference for sets, functions, relations and their applications is Birkhoff and Bartree [5].

II Naive category theory

Our objective in this chapter is to assemble some basics of the theory and to acquire some practice with the language of categories. There are several very good introductory texts for mathematicians and the ones by MacLane [67] and Herrlich and Strecker [46] are probably the best references for physicists who want more details than we can provide here. Note that we use the term set where MacLane uses the term small set: to mean a class that is not a proper class (cf. [46]). The important technical point is that we must not speak of the 'set of all sets' or 'the set of all groups' these are not constructible from the operations of standard set theory; they are examples of proper classes. We shall suppose that any group, ring, field, vector space or topological space is a set together with some extra structure, that is a class which is not a proper class. Occasionally we shall need to use maps between classes, then we use the same notation as for maps between sets.

1. CATEGORIES AS STRUCTURED GRAPHS

After the natural number system it could be argued that the most widely applied concept from mathematics is that of a graph. In physics we are concerned with the representation of observable phenomena and this is in principle possible by means of finite graphs. The vertices are observable entities of interest and the edges can represent interactions or motions that may be qualitative or, at best, quantitative with finite precision. When we come to fitting observations in a behavioural model the method is to cloak their graphical skeleton, usually with continuum theories because at most levels of experimentation real phenomena seem to suffer small changes when the ambient conditions undergo small changes. (The classical exceptions of course are changes of state and fracture phenomena where a continuous input of energy can yield a discontinuous change in behaviour. Such exceptional behaviour, and not only from physics, is currently an active area of application for catastrophe theory, cf. Poston and Stewart [79]). So for much of theoretical physics, topology, the study of continuous processes, is pre-eminent. However, the graphs come back again in the form of superstructure, such as group symmetries, on the topological spaces. In mathematics, the concept of graph enjoys ubiquity. The facet which we use in the sequel is category theory, which allows us to abstract similarities in the application of different mathematical constructions and to exploit a powerful common language. The increasing sophistication of studies in spacetime geometry and associated field theories using, for example, bundle theory and algebraic topology/geometry gradually is feeding into physics some of the categorical language. Inevitably, more category theory will be used at the formulation stage of physical theories and, there, care will be needed to avoid unnecessary obscurity so in this section we collect some basic terms, with examples, intended to give an accurate though not detailed impression of the ingredients of category theory. Subsequently we shall consider some commonly occuring construction processes that are essentially categorical.

1.1 Graph definitions

A graph is a collection $\boxdot$ of objects and a collection $\boxed{\uparrow\!\!\uparrow}$ of arrows with two operations, associating with each arrow a unique pair of objects, dom(a) and cod(a); that is to say,

$$\text{dom} : \boxed{\uparrow\!\!\uparrow} \to \boxdot, \quad \text{cod} : \boxed{\uparrow\!\!\uparrow} \to \boxdot ,$$

are maps. Actually this is a directed graph (cf. Higgins [47]) but we shall need no other kind. We wish to allow collections to be sets or proper classes (cf. [46]).

Example

The prototype of all graphs is that for transport systems, such as the London Underground. Then ⊡ is the set of stations and ⍐ is the set of train routes between stations. Two elements of ⊡ are Whitechapel and Stepney Green and there are just two arrows from the first to the second. These arrows are m_o (for Metropolitan Line) and d_o (for District Line) so we represent this situation locally by the diagram

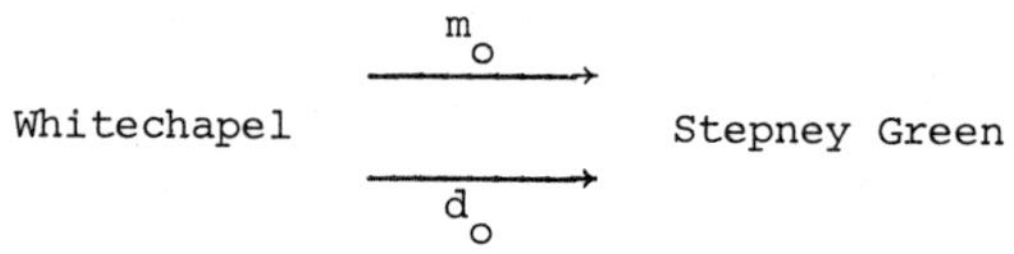

Here we have $\mathrm{dom}(m_o) = \mathrm{dom}(d_o) =$ Whitechapel

$\mathrm{cod}(m_o) = \mathrm{cod}(d_o) =$ Stepney Green

There are of course many more ways by Underground from Whitechapel to Stepney Green. The next shortest after the above pair is a composition of five arrows $m_3 \circ c_2 \circ c_1 \circ m_2 \circ m_1$:

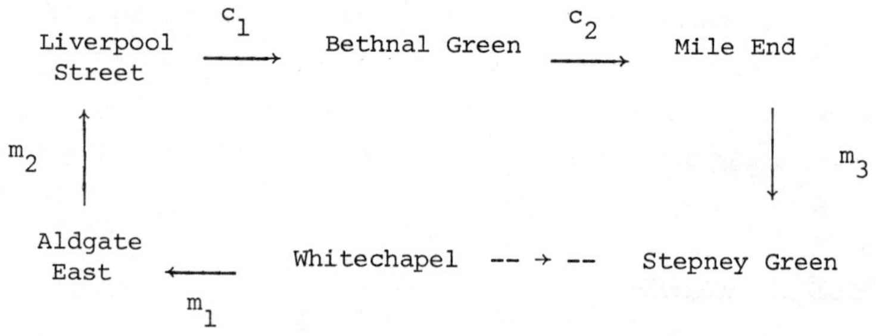

London Transport calls this composite way an available route and so we could include it in the set ⍐ . Plainly, we can only compose contiguous arrows and since the idea of Underground maps is to help the traveller, it is clearer to leave the traveller to do his own compositions. Hence, route maps only show the normally nonstopping routes between stations.

1.2 *Category axioms*

A category is a graph in which :

(i) every object A has an identity arrow I_A with $\text{dom}\ I_A = \text{cod}\ I_A = A$;

(ii) there is the possibility of composition of arrows in the sense that given any arrows $A \xrightarrow{f} B$ and $C \xrightarrow{g} D$ then they compose to give a unique arrow $A \xrightarrow{g \circ f} D$ if and only if $B = C$; if $B \neq C$ then $g \circ f$ is not defined;

(iii) wherever it is defined, composition is associative: $(g \circ f) \circ h = g \circ (f \circ h)$, (cf. §1.6 Ex.1);

(iv) identity arrows always compose with $A \xrightarrow{f} B$ to give $I_B \circ f = f$ and $f \circ I_A = f$, (cf.§1.6 Ex.2).

In a category, the arrows are usually called morphisms.

Remark

As a refinement of the concept of a graph these defining properties are fairly natural; associativity of set-theoretic maps prompts (iii). An elementary algebra with a non-associative composition is that of the Cayley numbers or octonions (cf. Porteous [78]) and we shall make use in the sequel of Lie algebras, which are also non-associative.

Example 1

The categories most commonly met in applications have objects that are structured sets and morphisms that are structure-preserving maps with the usual map composition. Among these are:

Grp, consisting of groups and group homomorphisms;

Ab , consisting of Abelian groups and group homomorphisms;

Vec, consisting of vector spaces and linear maps;

Top, consisting of topological spaces and continuous maps.

Example 2

A group $(G, \circ)$ is a category with one object (the underlying set), and each element is an arrow (actually a map from the set of elements to itself); also each arrow is invertible:

$$(\forall x \in G)(\exists x^{-1} \in G) \; : \; x \circ x^{-1} = x^{-1} \circ x = e = I_G.$$

Example 3

A category in which every arrow is invertible, that is :

$$(\forall\ A \xrightarrow{f} B)(\exists\ B \xrightarrow{f^{-1}} A) \; : \; f \circ f^{-1} = I_B,\ f^{-1} \circ f = I_A,$$

is called a groupoid. These are studied in detail by Higgins [47].

Example 4

Set is the category whose objects are all sets and whose arrows are all maps among them. This is an example of a large category: its collection of objects and arrows is not a set, it is a proper class. A category is called small if it is not large.

Example 5

We can make a larger category than Set by using the same objects but allowing all binary relations to be morphisms. This category is denoted Rel.

Example 6

Given a category $\mathcal{C}$ the dual or opposite category $\mathcal{C}^{op}$ is defined to have the same objects as $\mathcal{C}$ but for each morphism

$$A \xrightarrow{f} B \quad \text{in } \mathcal{C}$$

we get an opposite morphism, by interchanging the dom and cod maps,

$$A \xleftarrow{f^{op}} B \quad \text{in } \mathcal{C}^{op}$$

and composition in $\mathcal{C}^{op}$ is defined by

$$f^{op} \circ g^{op} = (g \circ f)^{op}. \qquad \text{(cf. §1.4 Ex.1)}$$

Example 7

Given categories $\mathcal{C}_1$, $\mathcal{C}_2$ their product category $\mathcal{C}_1 \times \mathcal{C}_2$ has as objects all ordered pairs of objects from $\mathcal{C}_1$ and $\mathcal{C}_2$ and morphisms are ordered pairs of morphisms from $\mathcal{C}_1$ and $\mathcal{C}_2$.

1.3   *Covariant functor*

A covariant functor F from a category C_1 to a category C_2 is
a pair of maps

$$F: \begin{cases} F_{obj} & : \text{ Objects of } C_1 \longrightarrow \text{ Objects of } C_2 \\ \\ F_{mor} & : \text{ Morphisms of } C_1 \longrightarrow \text{ Morphisms of } C_2 \end{cases}$$

sending

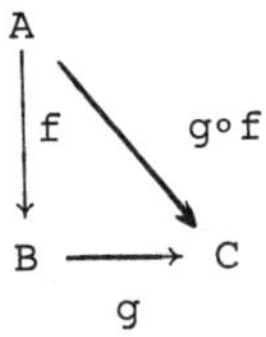

and respecting composition and identities :

$$F_{mor}(g \circ f) = F_{mor}(g) \circ F_{mor}(f)$$

$$F_{mor}(I_A) = I_{F_{obj}(A)}$$

Usually we just refer to covariant functors as <u>functors</u>, and also
we omit the subscripts from F_{obj} and F_{mor} since it should be
clear whether we are acting on an object or a morphism.   The
functor F is called :   <u>faithful</u> if F_{mor} is injective; <u>full</u>
when F_{mor} is surjective onto the C_2-morphisms among objects in
the image of F_{obj}; an <u>isomorphism</u> if both F_{obj} and F_{mor} are
bijective.

Example 1

The power set functor from Set to Set is defined by

$$P : f \begin{array}{c} A \\ \downarrow \\ B \end{array} \longmapsto \begin{array}{c} \text{Sub } A \\ \downarrow f_{\rightarrow} \\ \text{Sub } B \end{array}$$

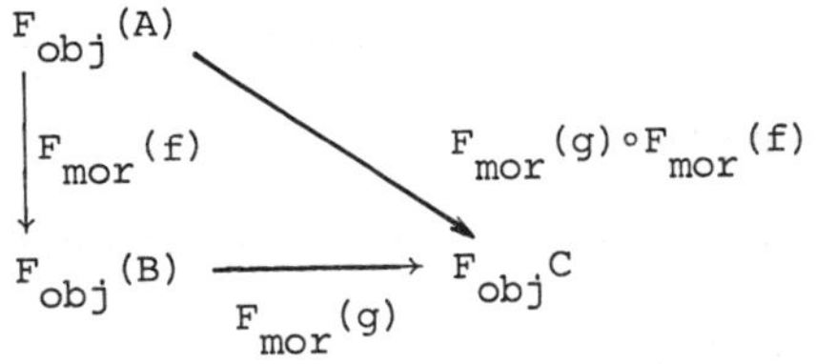

where,

$$f_{\rightarrow} : \text{Sub } A \longrightarrow \text{Sub } B : S \longmapsto \{f(x) \in B \mid x \in S\}.$$

It is customary to omit the subscript from $f_{\to}$.

Example 2

A category C is called concrete if there is a faithful functor F from C to Set, then F is called a forgetful functor, since it simply forgets the extra structure in C. Similarly, forgetful functors arise between other categories.

Example 3

Composites of full/faithful functors are also full/faithful functors.

Example 4

The categories Grp, Ab, Vec and Top are concrete categories.

Example 5

The set of all (small) categories with functors among them forms a large category, Cat.

1.4 Contravariant functor

A contravariant functor F from a category C_1 to a category C_2 is a pair of maps

$$F : \begin{cases} F_{obj} : \text{Objects of } C_1 \longrightarrow \text{Objects of } C_2 \\ F_{mor} : \text{Morphisms of } C_1 \longrightarrow \text{Morphisms of } C_2 \end{cases}$$

sending

$$\begin{array}{c} A \\ \downarrow f \\ B \end{array} \quad \text{to} \quad \begin{array}{c} F_{obj}(A) \\ \uparrow F_{mor}(f) \\ F_{obj}(B) \end{array}$$

in C_1 in C_2

and respecting identities but reversing compositions :

$$F_{mor}(g \circ f) = F_{mor}(f) \circ F_{mor}(g)$$

$$F_{mor}(I_A) = I_{F_{obj}(A)} .$$

We do not drop the qualification 'contravariant' when referring to these entities, though it is usual to drop the subscripts to the maps F_{mor}, F_{obj}.

Remark

The origin of the term contravariant is plain: we see that, in the image, morphisms are reversed with respect to objects.

Example 1

The dualising operation (§1.2 Ex.6) is a contravariant functor from any category to its opposite.

Example 2

Real vector spaces and linear maps among them form a large category. $\mathbb{R}$-Vec. Every vector space V in this category has a dual space $V^* = L(V;\mathbb{R})$, consisting of real valued linear maps from V to $\mathbb{R}$. Also, for every morphism $f : V \longrightarrow W$ in $\mathbb{R}$-Vec there is a dual linear map

$$f^* : W^* \longrightarrow V^* : x \longmapsto x \circ f.$$

Hence we obtain a contravariant functor from $\mathbb{R}$-Vec to itself by taking duals :

$$\begin{array}{ccc} V & & V^* \\ f\big\downarrow & \longmapsto & \big\uparrow f^* \\ W & & W^* \end{array}$$

Evidently $\mathbb{R}$ can be replaced by any field and we could equally well work with modules over any ring.

Example 3

There is a <u>contravariant power set functor</u> from Set to Set given by

$$P^* : \begin{array}{ccc} A & & \text{sub } A \\ f\big\downarrow & \longmapsto & \big\uparrow \overleftarrow{f} \\ B & & \text{sub } B \end{array}$$

where, as before, $\overleftarrow{f}$ is the inverse image map on subsets:

$$\overleftarrow{f} : \text{sub } B \longrightarrow \text{sub } A : S \longmapsto \{x \in A \mid f(x) \in S\}.$$

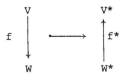

As with the term contravariant, we cannot afford to omit the superscript from $\overleftarrow{f}$, it reminds us of the backwards going effect of contravariance on morphisms.

1.5 *Subcategory*

We call a category $\mathcal{C}_o$ a subcategory of a category $\mathcal{C}$ if the objects and morphisms of $\mathcal{C}_o$ form a subclass of those of $\mathcal{C}$ and the inclusion maps

$$F_{obj} : \text{Objects of } \mathcal{C}_o \hookrightarrow \text{Objects of } \mathcal{C} : A \longmapsto A$$

$$F_{mor} : \text{Morphisms of } \mathcal{C}_o \hookrightarrow \text{Morphisms of } \mathcal{C} : f \longmapsto f$$

form a functor, the inclusion functor. We say that $\mathcal{C}_o$ is a full subcategory of $\mathcal{C}$ when the inclusion functor exists and is full.

Example 1

The category of all finite sets and maps among them is a full subcategory of Set; the category of all Hausdorff spaces is a full subcategory of the (large) category Top of topological spaces with continuous maps among them.

Example 2

The category Ab of all Abelian groups and homomorphisms among them is a (full) subcategory of Grp the (large) category of all groups and homomorphisms among them.

Example 3

Topological metric spaces and isometries (distance preserving maps) among them form a subcategory of Top. However, the inclusion is not full because even though every topological space can be given at least the trivial metric

$$d(x,y) = 1 \text{ if } x \neq y, \quad d(x,x) = 0 ,$$

not every continuous map need be an isometry in any available metric.

1.6 *Diagram*

A diagram is rather tricky to define, because of the freedom we normally wish to exercise in viewing as equivalent, different representations of a given collection of objects and morphisms. For convenience we use :

A diagram in a category $\mathcal{C}$ is a subclass Δ of $\mathcal{C}$ with specified objects $\{A_i | i \in J_o\}$ and morphisms

$$\{a_{ij}^k : A_i \longrightarrow A_j | k \in J_m, \ i, \ j \in J_o\}$$

for some indexing sets J_o and J_m, $J_o \neq \emptyset$. To avoid technical difficulties, we shall agree to interpret this definition with a spirit of reasonableness.

Remark

Most of the applications of category theory are concerned with diagrams, rather than with the whole category or subcategories. Much of our work will be concerned with diagrams from Top and Set. A diagram is called commutative if whenever two morphisms are obtained by composing morphisms in the diagram along different routes between the same endpoints, the two morphisms are equal.

Example 1

The associative law of composition in any category (cf. §1.2. iii) is equivalent to the requirement that the following diagram is commutative whenever the constituent compositions are defined.

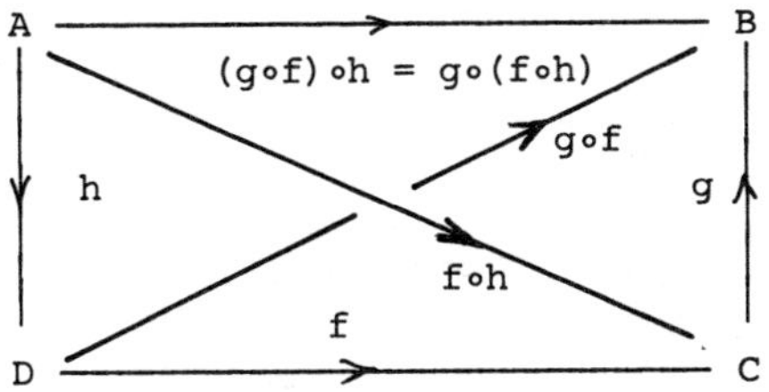

Example 2

The composition rule for identity morphisms in any category (cf. §1.2. iv) is equivalent to the requirement that the following diagram is commutative.

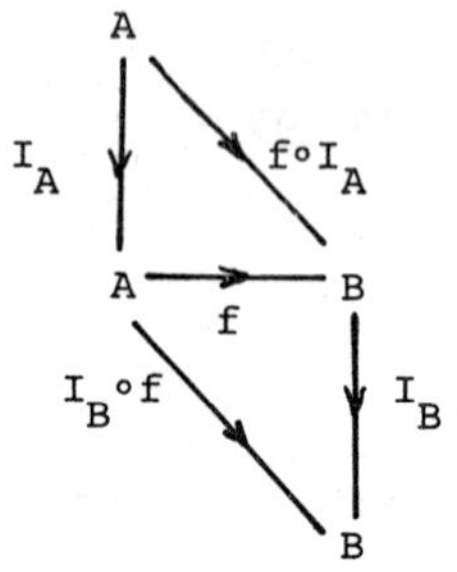

<u>Example 3</u>

The diagram (in Set, Top and Grp)

$$\Delta \;=\; \{\mathbb{R} \overset{a}{\underset{b}{\rightrightarrows}} \mathbb{R} \,|\, a : x \longmapsto 2x,\ b : x \longmapsto 3x\}$$

is plainly not commutative.

<u>Remark</u>

We shall henceforth use only diagrams that are <u>small</u> in the technical sense : the collections of objects and morphisms in a diagram will always belong to some fixed universe of sets. This does not restrict us to finite or even countable diagrams, of course. Diagrams crop up naturally in physics and much theory is devoted to their interpretation. Category theory provides, among other things, an abstract calculus for diagram chasing and in particular it prescribes certain natural objects and morphisms that may be available, by standard construction, for incorporation in given diagrams.

1.7 <u>Natural transformation</u>

Given two (covariant) functors F, H from $\mathcal{C}_1$ to $\mathcal{C}_2$ then a <u>natural transformation</u> from F to G is denoted by $\tau : F \stackrel{\bullet}{\rightarrow} G$ and defined to be a map

$$\tau : \text{Objects of } \mathcal{C}_1 \longrightarrow \text{Morphisms of } \mathcal{C}_2 : A \longmapsto \tau(A)$$

such that every diagram

$$\begin{array}{c} A_1 \\ a \big\downarrow \\ A_2 \end{array} \quad \text{in } \mathcal{C}_1$$

yields in $\mathcal{C}_2$ the following commutative diagram.

$$\begin{array}{ccc} F(A_1) & \xrightarrow{\tau(A_1)} & G(A_1) \\ F(a) \big\downarrow & & \big\downarrow G(a) \\ F(A_2) & \xrightarrow{\tau(A_2)} & G(A_2) \end{array}$$

A natural transformation $\tau : F \overset{\bullet}{\to} G$ is called a <u>natural equivalence</u> or <u>natural isomorphism</u> if

$$\left(\forall \tau(A)\right)\left(\exists \tau(A)^{-1}\right) : \begin{cases} \tau(A) \circ \tau(A)^{-1} = I_{G(A)} \\ \tau(A)^{-1} \circ \tau(A) = I_{F(A)} \end{cases}$$

in that case F and G are called <u>naturally isomorphic functors</u>.

<u>Example 1</u>

Natural transformations compose to give a natural transformation. Hence we can obtain $\mathcal{C}_2^{\mathcal{C}_1}$ a <u>functor category</u> whose morphisms are natural transformations and whose objects are functors from $\mathcal{C}_1$ to $\mathcal{C}_2$. The objects of $\mathcal{C}_1^{\mathcal{C}_2^{op}}$ are contravariant functors from $\mathcal{C}_2$ to $\mathcal{C}_1$ (cf. §1.2 Ex.6).

<u>Example 2</u>

For any functor F there is a natural isomorphism $1_F : F \overset{\bullet}{\to} F$ defined by $1_F(A) = I_{F(A)}$.

<u>Example 3</u>

For finite-dimensional vector spaces we have the most celebrated natural isomorphism, the <u>taking of double duals</u>. Here, F is the identity functor and

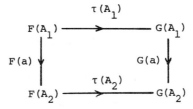

$$G : \begin{array}{ccc} V & & V^{**} \\ f \downarrow & \longmapsto & \downarrow f^{**} \\ W & & W^{**} \end{array}$$

where (cf. §1.4. Ex. 1)

$$V^{**} = (V^*)^* = L(V^*;\mathbb{R})$$

and the constituent maps of the functor G are given by

$$\begin{cases} V \longrightarrow V^{**} : x \longmapsto x^{**} \begin{cases} x^{**} : V^* \longrightarrow \mathbb{R} \\ \quad\;\; : \alpha \longmapsto \alpha(x) \end{cases} \\ L(V;W) \longrightarrow L(V^{**};W^{**}) : f \longmapsto f^{**} \begin{cases} f^{**} : V^{**} \longrightarrow W^{**} \\ \quad\;\; : a \longmapsto a \circ f^*. \end{cases} \end{cases}$$

The importance of these maps lies in their freedom from arbitrary choices; for example, they do not depend on what bases are used in the spaces. This freedom is what is referred to as naturality in their construction; it is characteristic of natural isomorphisms. In physical theories, often the term canonical is used with a similar meaning.

1.8 Special morphisms

In a category C a morphism $f : A \longrightarrow B$ is called :

(i) invertible in C if $\exists g : B \longrightarrow A$ in C with $g \circ f = I_A$ and $f \circ g = I_B$, then we write $A \cong B$ and say that A and B are isomorphic and f is an isomorphism in C;

(ii) monic (or a monomorphism) in C if whenever there are

$$k,\ell : C \longrightarrow A \qquad \text{with } f \circ k = f \circ \ell$$

then $k = \ell$;

(iii) epic (or an epimorphism) in C if whenever there are

$$k,\ell : B \longrightarrow D \qquad \text{with } k \circ f = \ell \circ f$$

then $k = \ell$.

Example 1

In Set the monics are the injections (one-to-one non-empty maps) the epimorphisms are the surjections (onto maps) and the isomorphisms are the bijections (one-to-one and onto maps.)

Example 2

In Top, the morphisms are continuous maps and isomorphisms are homeomorphisms. It is well known that though every continuous bijection has an inverse map, it need not be continuous (try mapping an interval round a circle) and hence need not be invertible in Top, though it is invertible in Set.

Example 3

In the category Rng of rings and ring maps, the inclusion map of the integers in the rationals is epic and monic but not an isomorphism.

1.9 *Special Objects*

In a category $\mathcal{C}$ an object A is called

(i) terminal if to each object B in $\mathcal{C}$ there is exactly one morphism $B \longrightarrow A$;

(ii) initial if to each object B in $\mathcal{C}$ there is exactly one morphism $A \longrightarrow B$

(iii) null or zero if it is both terminal and initial.

Remark

I_A is the only morphism $A \longrightarrow A$ if A is terminal; also, any two terminal objects are isomorphic and any two initial objects are isomorphic.

Example 1

In Set (and Top), any one point set $\{x\}$ is a terminal object, and the empty set is an initial object (the empty map is the unique morphism from the empty set to any set).

Example 2

In Grp, the singleton group $(\{e\}, \circ)$ is a null object; similarly the trivial vector space $\{\underline{0}\}$ in Vec.

Example 3

The ring of integers $(\mathbb{Z}, +, \cdot)$ is an initial object in the category of rings and ring-maps and the singleton rings like $(\{0\}, +, \cdot)$ are terminal objects.

Example 4

The category of fields and field-maps has no initial object and no terminal object (recall that the smallest field has two elements, 0 and 1, but in a ring we may have $0 = 1$).

1.10 Inverse morphisms

Given a morphism $f : A \longrightarrow B$ in a category $\mathcal{C}$,

(i) a right inverse or section or coretraction of f is a morphism $g : B \longrightarrow A$ with $f \circ g = I_B$;

(ii) a left inverse or retraction for f is a morphism $h : B \longrightarrow A$ with $h \circ f = I_A$.

Remark

If f has a section g then f is epic (cf. §1.8 iii). For suppose also $k, \ell : B \longrightarrow D$ and $k \circ f = \ell \circ f$; then $k \circ f \circ g = k \circ I_B = k = \ell \circ f \circ g = \ell \circ I_B = \ell$. The assertion that every set-theoretic surjection has a section is called the axiom of choice in set theory; it is equivalent to Zorn's Lemma. If f has a retraction h then f is monic (cf. §1.8 ii). For suppose also $k, \ell : C \longrightarrow A$ and $f \circ k = f \circ \ell$; then $h \circ f \circ k = I_A \circ k = k = h \circ f \circ \ell = I_A \circ \ell = \ell$.

Example 1

In Set a morphism is a section if and only if it is injective and not the empty map from $\emptyset$ to a non-empty set; but in Grp this condition is necessary though not sufficient.

Example 2

In Top, $g : B \longrightarrow A$ is a section of $f : A \longrightarrow B$ if and only if g is a continuous injection, continuously invertible on $g(B) \subseteq A$.

Example 3

In Set a morphism is a retraction if and only if it is surjective.

Example 4

A tangent vector field on a smooth manifold M is a section $V : M \longrightarrow TM$ of the canonical surjection $\Pi : TM \longrightarrow M$. This simply means that $\Pi \circ V = I_M$ and so we know that V attaches to each point $x \in M$ a vector $V(x)$ from $T_x M$, the tangent vector space to M at x.

2. STRUCTURES ON CATEGORIES

We have in the previous section the basic terms in a language that exploits commonly occurring patterns in mathematical constructions. In applications to physical theories these same distinctive patterns have significant interpretations and often the interplay of different branches of mathematics can be viewed through the roles of functors. Frequently, the objects (e.g., geometrical or configuration spaces) in physical theories are derived from more primitive objects (usually potentially observable interactions) and a similar thing happens in mathematics. It turns out that many of these processes of derivation are examples of the taking of limits of diagrams in a category. Closely related to natural constructions being carried from one category to another is the concept of adjoint functors. In this section we shall pursue these two ideas.

2.1 *Limits of diagrams*

Denote by Δ the diagram (cf. §1.6)

$$\{a^k_{ij} : A_i \longrightarrow A_j \mid k \in J_m;\ i,\ j \in J_o\}$$

in a category $\mathcal{C}$, for some indexing sets J_m, J_o. We define as follows:

(i) A left limit (if it exists) of Δ is an object L and morphisms $\{f_i : L \longrightarrow A_i \mid i \in J_o\}$ in C, commuting with the a^k_{ij} and such that if $\{g_i : K \longrightarrow A_i \mid i \in J_o\}$ in C also commutes with the a^k_{ij} then there exists a unique morphism $\ell : K \longrightarrow L$ such that $(\forall i \in J_o)\ g_i = f_i \circ \ell$. If L is such a left limit object of Δ then we sometimes write (loosely) $L = \varprojlim \Delta$, the existence of the morphisms, and ℓ with the universal property, being understood. We can speak of the left limit because, if it exists, it is determined up to isomorphism in C.

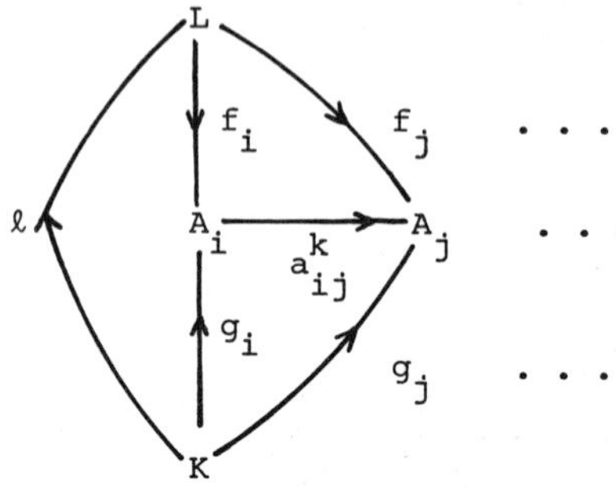

Remark

Left limits are sometimes called just limits or projective limits or left roots, or inverse limits.

(ii) A right limit (if it exists) of Δ is an object R and morphisms $\{f_i : A_i \longrightarrow R \mid i \in J_o\}$ in C, commuting with the a^k_{ij} and such that if $\{g_i : A_i \longrightarrow K \mid i \in J_o\}$ in C also commutes with the a^k_{ij} then there exists a unique morphism $r : R \longrightarrow K$ such that $(\forall i \in J_o)\ g_i = r \circ f_i$.

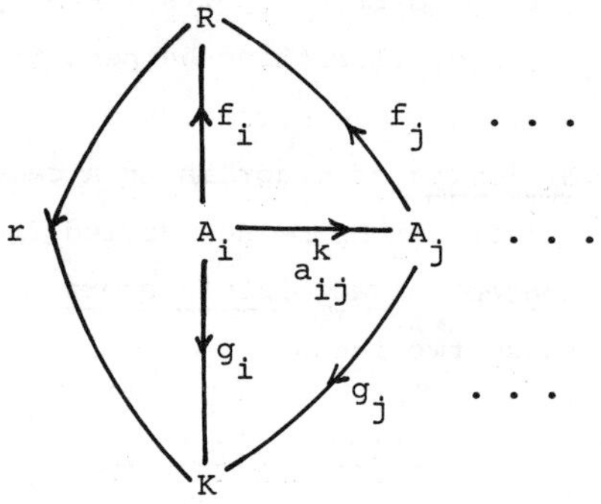

Then, as before, such a limit is determined up to isomorphism in $\mathcal{C}$, and we write here $\varinjlim \Delta = R$.

Remark

Right limits are sometimes called <u>colimits</u>, <u>inductive limits</u> or <u>right roots</u> or <u>direct limits</u>.

Example 1

Consider in Set any diagram like $\{A_1 \xleftarrow{a_1} A_o \xrightarrow{a_2} A_2\}$.
This always has left limit object

$$L = A_o \quad \text{with} \quad f_i = a_i \quad \text{for} \quad i = 1, 2 \text{ and } f_o = I_{A_o}.$$

For suppose we have

$\{g_i : K \longrightarrow A_i | i=0,1,2\}$

commuting with the a_i, then $g_o : K \longrightarrow A_o$ is the required unique map with

$g_i = f_i \circ g_o$.

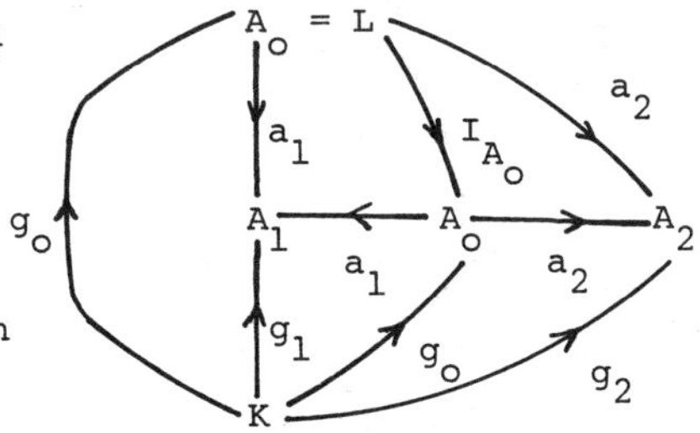

(Observe that if we omit A_1 and a_1 from our diagram the left limit is unchanged.)

Example 2

Consider, in Set, any diagram like $\{A_1 \xrightarrow{a_1} A_o \xleftarrow{a_2} A_2\}$.
This always has right limit object $R = A_o$ with $f_i = a_i$ for $i = 1, 2$ and $f_o = I_{A_o}$.

For suppose we have

$\{g_i : A_i \longrightarrow K | i = 0,1,2\}$

commuting with the a_i, then $g_o : A_o \longrightarrow K$ is the required unique map with $g_i = g \circ f$. (If we omit A_2, a_2 from our diagram the right limit is unchanged.)

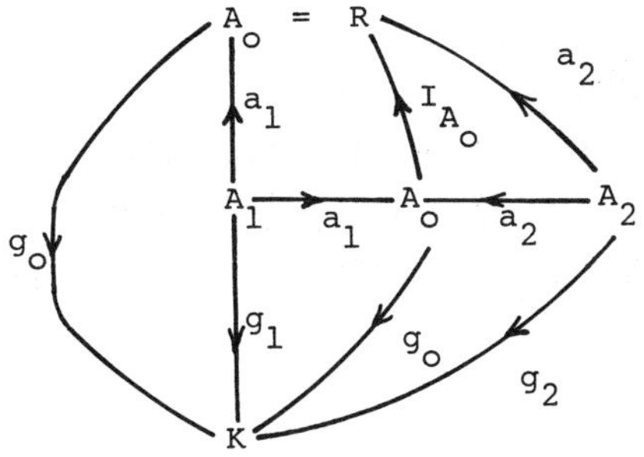

Example 3

Consider in Set the diagram $\{A_1, A_2\}$, consisting of two sets and no morphisms.

The left limit is $A_1 \times A_2$, the Cartesian product consisting of all ordered pairs of elements from A_1 and A_2. ($A_1 \times A_2 = \emptyset$ if at least one of A_1, A_2 is $\emptyset$.) In this case each f_i is projection onto the i-th component and the map

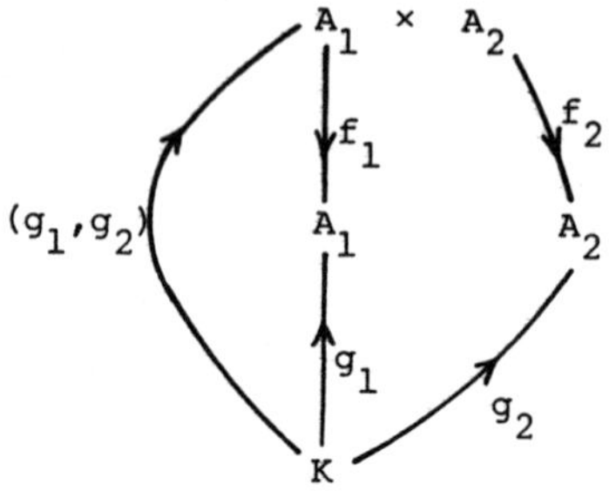

$$(g_1, g_2) : K \longrightarrow A_1 \times A_2 : x \longmapsto (g_1(x), g_2(x))$$

is evidently unique.

Example 4

Consider again in Set the diagram $\{A_1, A_2\}$, consisting of two sets and no morphisms. It has right limit $A_1 \amalg A_2$, the disjoint union of A_1 and A_2, which is the union of disjoint copies of A_1 and A_2.

There are obvious injections $f_i : A_i \longrightarrow A_1 \amalg A_2$ and the required map $r : A_1 \amalg A_2 \longrightarrow K$ is uniquely determined by $r : x \longmapsto g_i(x)$ if $x \in \overset{\leftarrow}{f_i} A_i$.

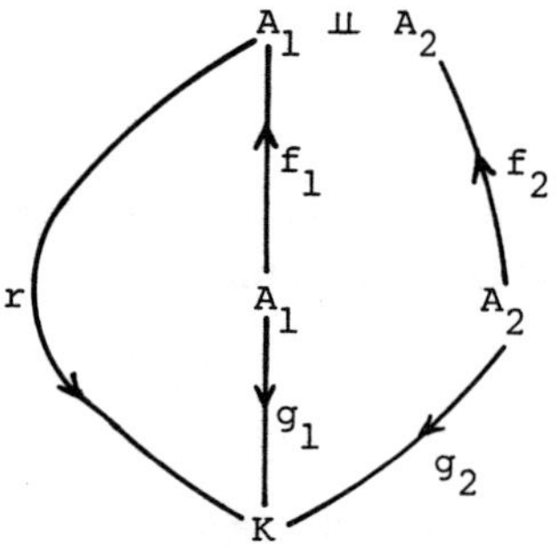

Remark

We shall encounter these four examples again to find the other limits in Examples 1, 2 but first we generalize 3, 4 to arbitrary collections of sets.

2.2 *Products and coproducts*

Let J be an index set and consider a diagram $\Delta_o = \{A_i | i \in J\}$ with no morphisms, in a category $\mathcal{C}$.

(i) If it exists, the left limit object $\underleftarrow{\mathrm{Lim}}\, \Delta_o$ is called the product in $\mathcal{C}$ of the objects $\{A_i | i \in J\}$; it is denoted by $\prod_{i \in J} A_i$ and the associated maps $\pi_j : \prod_{i \in J} A_i \longrightarrow A_j$ are called projections.

(ii) If it exists, the right limit object $\underrightarrow{\mathrm{Lim}}\, \Delta_o$ is called the coproduct in $\mathcal{C}$ of the objects $\{A_i | i \in J\}$; it is denoted by $\coprod_{i \in J} A_i$ and the associated maps $\mu_j : A_j \longrightarrow \coprod_{i \in J} A_i$ are called injections.

Example 1

As we expect from §2.1 Ex. 3, in Set the product of objects $\{A_i | i \in J\}$ does exist, it is their Cartesian product with projections $\pi_j : \prod_{i \in J} A_i \longrightarrow A_j : (x_i)_{i \in J} \longmapsto x_j$.

Example 2

The product of objects $\{(A_i, \circ) | i \in J\}$ in Grp is the direct product group $(\prod_{i \in J} A_i, \circ)$ where $\prod_{i \in J} A_i$ is the set product and composition is

$$\left(a_i\right)_{i \in J} \circ \left(b_i\right)_{i \in J} = \left(a_i \circ b_i\right)_{i \in J} .$$

Example 3

The product of topological spaces $\{(A_i, T_i) | i \in J\}$ in Top (cf. III §1 below) again uses the set product and has the smallest topology (cf. III §2 below) that makes the projections continuous. So $\prod_{i \in J} A_i$ has as open sets only those U such that for some $U_i \in T_i$ (i.e., U_i is open in A_i) $\pi_i U = U_i$. Hence U must be of the form $\bigcap \overleftarrow{\pi_i} U_i$, for some open $U_i \subseteq A_i$.

Example 4

Further to §1.2 Ex. 4, in Set the coproduct of objects $\{A_i | i \in J\}$ does exist, it is their disjoint union, $\coprod_{i \in J} A_i$ with injections

$$f_j : A_j \longrightarrow \coprod_{i \in J} A_i : x \longmapsto x ,$$

and any collection of maps

$$\{g_j : A_j \longrightarrow K \mid j \in J_o\}$$

determines a unique map

$$r : \coprod_{i \in J} A_i \longrightarrow K : (x \in A_j) \longmapsto g_j(x) .$$

Example 5

Coproduct carries over to Grp, to give the free product group of a collection of groups, using the disjoint operations on the disjoint union. In Top we give $\coprod_{i \in J} A_i$ the largest topology that gives continuity to the injections, so $U \subseteq \coprod_{i \in J} A_i$ is open only if for all $i \in J$ we can find open $U_i \subseteq A_i$ with $\overleftarrow{f_i}U = U_i$. The topological space so formed is called the topological disjoint sum of the spaces involved.

Example 6

In the case that J is a singleton, then the diagram consists of one object and it is its own left and right limit. To see this, omit A_1, A_2 and a_1, a_2 from the diagram in §2.1 Ex.1.

2.3 *Pullback and pushout*

(i) In any category, given a diagram

$$A_1 \xrightarrow{a_1} A_o \xleftarrow{a_2} A_2$$

then, if it exists, its left limit is called the pullback (or fibred product of A_1 and A_2) over A_o and the completed diagram so formed is called a pullback square; it is commutative.

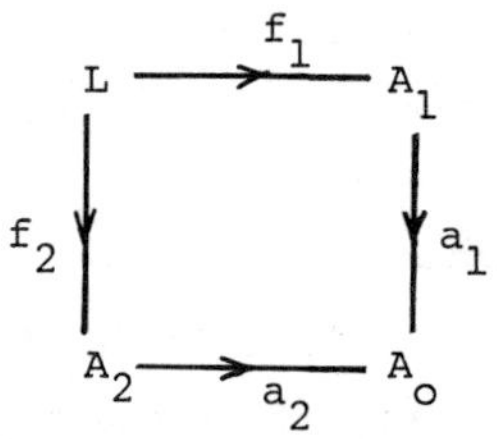

Remark

Either of $a_1 \circ f_1$, $a_2 \circ f_2$ will

serve as $f_o : L \longrightarrow A_o$ (cf. §2.1 i).

Example 1

In Set we construct the pullback as follows:

$$L = \coprod_{i=1,2} X_i \quad \text{with} \quad X_1 \subseteq A_1,\ X_2 \subseteq A_2$$

given by

$$X_1 = \{x \in A_1 \mid a_1(x) \in a_2 A_2\}$$

$$X_2 = \{y \in A_2 \mid a_2(y) \in a_1 A_1\} .$$

From the definition of left limit (cf. §2.1 i) we know that given any

$$A_1 \xleftarrow{g_1} K \xrightarrow{g_2} A_2$$

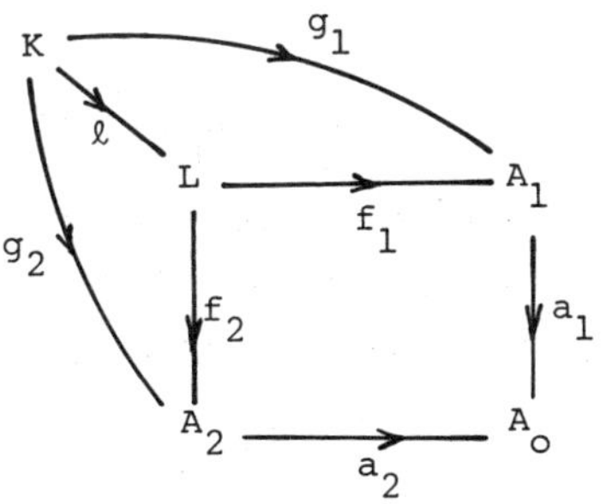

commuting with a_1, a_2, there is a unique morphism $\ell : K \longrightarrow L$ with $g_1 = f_1 \circ \ell$ and $g_2 = f_2 \circ \ell$. From §2.1 Ex.2 we know that the original diagram always has a right limit, namely A_o.

(ii) There is a dual to (i) : given a diagram

$$A_1 \xleftarrow{a_1} A_o \xrightarrow{a_2} A_2$$

in any category then, if it exists, its right limit is called the pushout (or fibred coproduct or fibred sum of A_1 and A_2) over A_o and the completed diagram so formed is called a pushout square ; it is commutative.

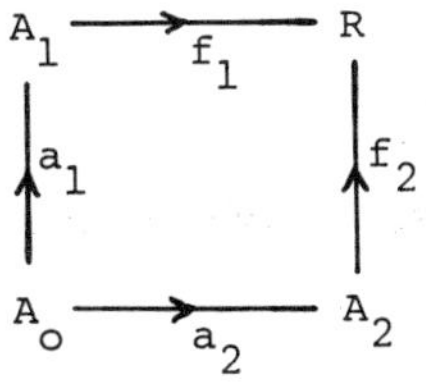

Remark

By commutativity we have

$$f_1 \circ a_1 = f_2 \circ a_2 = f_o : A_o \longrightarrow R .$$

Example 2

In Set we find R from an equivalence relation on $\coprod_{k=1,2} A_k$ defined by : $x \sim y$ if any of the following hold

(α) $x = y$

(β) $a_1(x) = y$ or $a_1(y) = x$

(γ) $a_2(x) = y$ or $a_2(y) = x$.

Then R is the set of equivalence classes

$$R = \{[x]_\sim | x \in \coprod_{k=1,2} A_k\}$$

where

$$[x]_\sim = \{y \in \coprod_{k=1,2} A_k | x \sim y\}$$

and the associated morphisms are (cf. §2.1 ii)
$f_i : A_i \longrightarrow R : x \longmapsto [x]_\sim$ sending elements to their equivalence classes. These morphisms are well-defined because we always look on the A_i as disjoint subsets of $\coprod_{k=1,2} A_k$.

It follows that given any

$$A_1 \xrightarrow{g_1} K \xleftarrow{g_2} A_2$$

commuting with a_1, a_2, there is a unique morphism $r : R \longrightarrow K$ such that $g_1 = r \circ a_1$ and $g_2 = r \circ a_2$. From §2.1 Ex. 1 we know that the original diagram always has A_o as left limit.

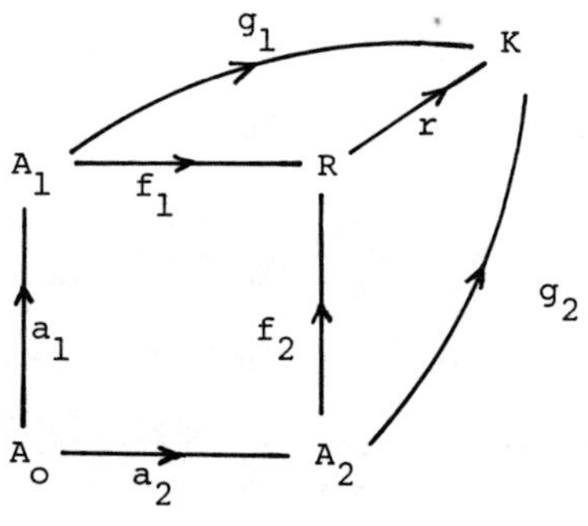

Example 3

In Set, given subsets A_1, A_2 of a set A_o we have inclusion maps a_1, a_2 :

$$A_1 \overset{a_1}{\hookrightarrow} A_o \overset{a_2}{\hookleftarrow} A_2$$

and the pullback object is the intersection $A_1 \cap A_2$.

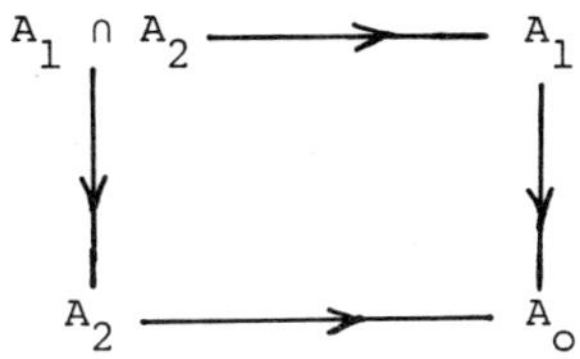

Example 4

The same construction as in Example 3 serves in Grp for subgroups A_1, A_2 of a group A_o (when necessarily the inclusion maps are group homomorphisms) and in Top for subsets (when necessarily the subspace topologies make the inclusion maps continuous).

Example 5

A typical situation that arises in bundle theory is the diagram

$$X \xrightarrow{\Pi_X} M \xleftarrow{\Pi_Y} Y$$

in Man, the category of smooth manifolds and smooth maps among them. Here X and Y are bundles over M with canonical projections Π_X, Π_Y onto M.
We obtain a pullback square with $L = \{(x,y) \in X \times Y \mid \Pi_X(x) = \Pi_Y(y)\}$ and the submanifold structure induced by inclusion in the product manifold X×Y. The smooth maps π_1, π_2 are just the restrictions of the usual projections from a Cartesian product set (cf. §2.2 Ex. 1).

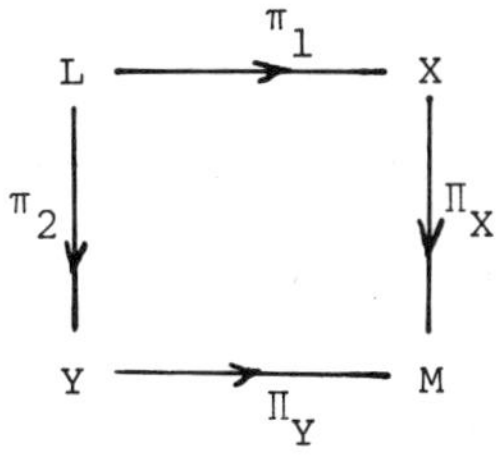

2.4 Equalizer and coequalizer

In any category, given a diagram

$$\Delta : A_1 \overset{a_1}{\underset{a_2}{\rightrightarrows}} A_2$$

(i) the equalizer is the left limit and

(ii) the coequalizer is the right limit, when they exist.

Example 1

In Set the equalizer object is

$L = \{x \in A_1 | a_1(x) = a_2(x)\}$

and the associated morphisms

$f_i : L \longrightarrow A_i$ are

inclusion in A_1 and its

composite with the restriction

of a_1 (or a_2) to $L \subseteq A_1$.

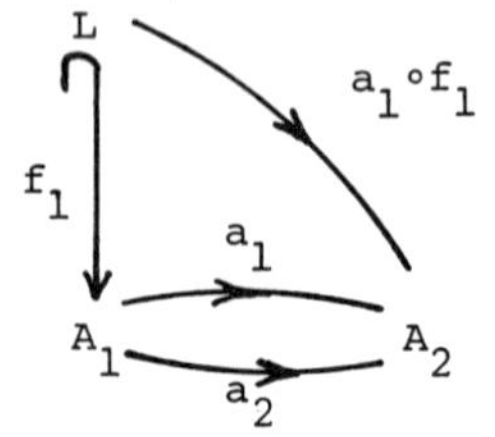

Example 2

The construction in the previous example carries over to Grp, Vec and Top by viewing L as a subobject of the object A_1 .

Example 3

In Set the coequalizer object is $R = A_2/\sim$ where $\sim$ is the smallest equivalence relation on A_2 that contains

$\{(a_1(x), a_2(x)) | x \in A_1\}$

and the associated morphisms are

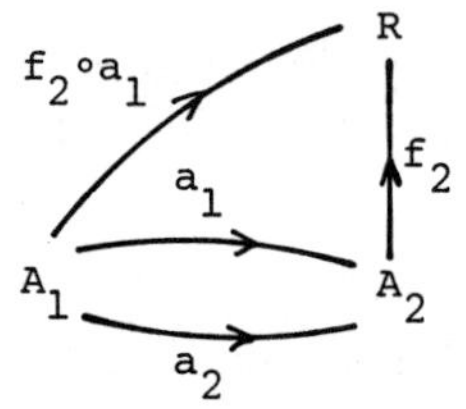

$f_2 : A_2 \longrightarrow R : x \longmapsto [x]_\sim$, $f_1 = f_2 \circ a_1$.

Example 4

In Top we use the coequalizer from Set with the quotient topology: $U \subseteq A_2/\sim$ is open if $\overleftarrow{f_2}U$ is open in A_2 . We also say that the topology for $A_2/\sim$ is coinduced by f_2 from A_2 , (cf. III §1).

2.5 *Complete category*

A category $\mathcal{C}$ is called

(i) left complete if every diagram has a left limit in $\mathcal{C}$,

(ii) right complete if every diagram has a right limit in $\mathcal{C}$,

(iii) complete if both left complete and right complete,

(iv) finitely left complete/right complete/complete if in (i), (ii), (iii) respectively we use only finite diagrams.

Warning

Some authors use complete and cocomplete instead of left complete and right complete.

Example 1

Some commonly occurring complete categories are : Set; Grp; Top; $\mathbb{R}$-Vec; Ab (Abelian groups); Rng (rings); R-Mod (modules over a ring R); Ab_{TF} (torsion free abelian groups); Cat; products of complete categories.

Example 2

The categories of finite sets and finite topological spaces are finitely complete but neither is complete. The category Met of metrizable topological spaces and continuous maps among them is countably left complete (all diagrams with countably many morphisms have left limits) but not complete.

Example 3

In a small category $\mathcal{C}$ (so the collections of objects and morphisms are sets) :

$$\mathcal{C} \text{ is left complete} \iff \mathcal{C} \text{ is right complete.}$$

Example 4

The category of finite groups is finitely left complete but not finitely right complete.

Example 5

The category of non-empty sets is neither finitely left nor finitely right complete.

Example 6

The category of fields and field maps is neither finitely left nor finitely right complete.

Example 7

A poset (partially ordered set) is a category and it is finitely complete if and only if it is a lattice with a smallest member and a largest member.

Example 8

The actual construction of the limits in Top and Grp relies heavily on that in Set. Also, Cartesian products and disjoint unions are basic constructions for any limits in Set (cf. §2.1 Exs. 3, 4 and §2.2 Exs. 1, 4). We can say that for any diagram

$$\Delta = \{a_{ij}^k : A_i \longrightarrow A_j \mid k \in J_m,\ i,j \in J_o\} \text{ in Set,}$$

(i) $\underleftarrow{\text{Lim}}\,\Delta$ is a subset $\prod_{i \in J_o} X_i$ of $\prod_{i \in J_o} A_i$ with $a_{ij}^k\, X_i = X_j$ for all $i,j \in J_o$, $k \in J_m$ and the restricted projections ;

(ii) $\underrightarrow{\text{Lim}}\,\Delta$ is the set of equivalence classes, $\{[x]_\sim \mid x \in \coprod_{i \in J_o} A_i\}$, of the smallest equivalence relation generated by

$$x \sim y \quad \text{if} \quad (\exists a_{ij}^k) : a_{ij}^k(x) = y ,$$

and the limit morphisms are the maps onto classes. (For a discussion and proof see Arbib and Manes [1] p.47.)

Example 9

A category is left complete if and only if it admits products and equalizers; it is right complete if and only if it admits coproducts and coequalizers. (For proof see Higgins [47] p.54.)

Example 10

All categories are <u>n-complete</u> for n = 1,2,3. By this we mean that in any category there are left and right limits for all diagrams like

A (cf. §2.2 Ex. 6)

$A \xrightarrow{a} B$ (cf. §2.1 Exs. 1 and 2, final observations)

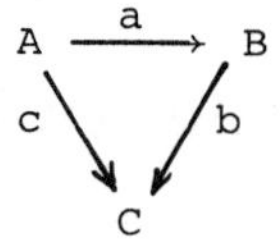

Example 11

For any category C the following are equivalent (cf. Herrlich and Strecker [46] p.158 for more) :

(i) $\mathcal{C}$ is finitely left complete;

(ii) $\mathcal{C}$ has (all) pullbacks and a terminal object;

(iii) $\mathcal{C}$ has (all) finite products and (all) pullbacks;

(iv) $\mathcal{C}$ has (all) finite products and equalizers.

This proposition, like others in category theory, has a dual form by changing terms as follows:

left complete	$\longrightarrow$	right complete
pullback	$\longrightarrow$	pushout
product	$\longrightarrow$	coproduct
equalizer	$\longrightarrow$	coequalizer

For more discussion of this duality principle see MacLane [67] p.31 et seq. also Herrlich and Strecker [46] throughout.

2.6 *Limit preserving functors*

Given a diagram Δ in a category $\mathcal{C}_1$ and a functor F from $\mathcal{C}_1$ to $\mathcal{C}_2$ we obtain a diagram in $\mathcal{C}_2$ which we can denote by $F(\Delta)$. Then we say :

(i) F preserves left limits if, for arbitrary diagrams Δ ,

$$(\exists \underleftarrow{\mathrm{Lim}}\, \Delta) \Rightarrow (\exists \underleftarrow{\mathrm{Lim}}\, F(\Delta)) = (F \underleftarrow{\mathrm{Lim}}\, \Delta) \ ;$$

(ii) F preserves right limits if, for arbitrary diagrams Δ ,

$$(\exists \underrightarrow{\mathrm{lim}}\, \Delta) \Rightarrow (\exists \underrightarrow{\mathrm{lim}}\, F(\Delta)) = (F \underrightarrow{\mathrm{lim}}\, \Delta) \ ;$$

(iii) F preserves limits or is continuous if it perserves both left and right limits.

Example 1

The forgetful functors from Grp and $\mathbb{R}$Mod to Set do not preserve right limits.

Example 2

The functor from Set to Ab, constructing free Abelian groups on sets does not preserve limits.

Example 3

The forgetful functor from Top to Set preserves limits.

Example 4

The inclusion functor Ab $\longrightarrow$ Grp preserves left limits but not right limits.

Example 5

The forgetful functor from the category of finite Abelian groups to Set preserves finite left limits but not arbitrary left limits.

2.7 *Functors of two variables*

We need to prepare for a definition in the next section by bringing together the notions of opposite category and product category (cf. §1.2 Exs. 6, 7).

Given three categories $\mathcal{C}_1$, $\mathcal{C}_2$ and $\mathcal{C}$, a functor

$$F : \mathcal{C}_1^{op} \times \mathcal{C}_2 \longrightarrow \mathcal{C}$$

is called a functor of two variables, contravariant in the first and covariant in the second.

Plainly the terminology extends easily to more variables, and for the physicist it is reminiscent of that in tensor analysis (cf. Dodson and Poston [28]).

Example 1

The illustration we need below is the standard case, for any category $\mathcal{C}$,

$$H_{\mathcal{C}} : \mathcal{C}^{op} \times \mathcal{C} \longrightarrow \text{Set}$$

and

$$H_{\mathcal{C}\text{obj}} : (A,B) \longmapsto \{\text{morphisms } A \longrightarrow B\}$$

$$H_{\mathcal{C}\text{mor}} : \begin{array}{cc} A_1 & B_1 \\ \uparrow f & \downarrow g \\ A_2 & B_2 \end{array} \longmapsto \begin{array}{c} \{\text{morphisms } A_1 \xrightarrow{h} B_1\} \\ \downarrow \\ \{\text{morphisms } A_2 \xrightarrow{g\circ h\circ f} B_2\} \end{array}$$

So what we are doing is to make this square diagram commute :

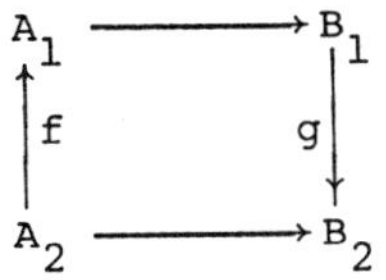

2.8 Adjoint functors

Certain functors crop up in pairs, one going each way, and they are important because they preserve limits (cf. §2.6) and because they characterise certain general constructions (cf. Higgins [47] p.15 et seq.).

Given two (covariant) functors

$$F : \mathcal{C}_1 \longrightarrow \mathcal{C}_2 \quad \text{and} \quad G : \mathcal{C}_2 \longrightarrow \mathcal{C}_1$$

we can by the construction in §2.6 obtain two functors

$$\tilde{F}, \tilde{G} : \mathcal{C}_1^{op} \times \mathcal{C}_2 \longrightarrow \text{Set}$$

by making compositions with $H_{\mathcal{C}_1}$, $H_{\mathcal{C}_2}$:

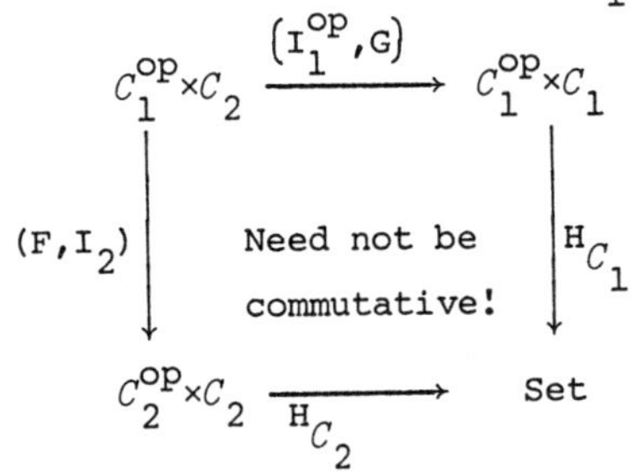

$$\begin{cases} \tilde{F} = H_{\mathcal{C}_2} \circ (F, I_2) \\ \tilde{G} = H_{\mathcal{C}_1} \circ (I_1^{op}, G) \end{cases}$$

In general this category diagram is not commutative. The next best thing is for the two composite functors $\tilde{F}$ and $\tilde{G}$ to be naturally equivalent (cf. §1.7); in that case we say that (F,G) is an <u>adjoint pair</u> of functors. We also say that F <u>is left adjoint to</u> G, or F <u>is a left adjoint</u> and F <u>has a right adjoint</u>. This relation between functors is not symmetric (cf. Ex. 5 below).

<u>Remark</u>

The intricacy of the definition somewhat obscures a fairly simple intuitive idea. Namely, that if (F,G) is an adjoint pair then:

(i) in $\mathcal{C}_1$, there are morphisms from objects in $\mathcal{C}_1$ to objects in $G_{obj}\mathcal{C}_2$ (e.g. $A \xrightarrow{f} G_{obj}B$) ;

(ii) in $\mathcal{C}_2$, there are morphisms from objects in $F_{obj}\mathcal{C}_1$ to objects in $\mathcal{C}_2$ (e.g. $F_{obj}A \xrightarrow{f^\dagger} B$) ;

(iii) there is a correspondence $f \longmapsto f^\dagger$ such as to make both of the following diagrams commute whenever one of them commutes.

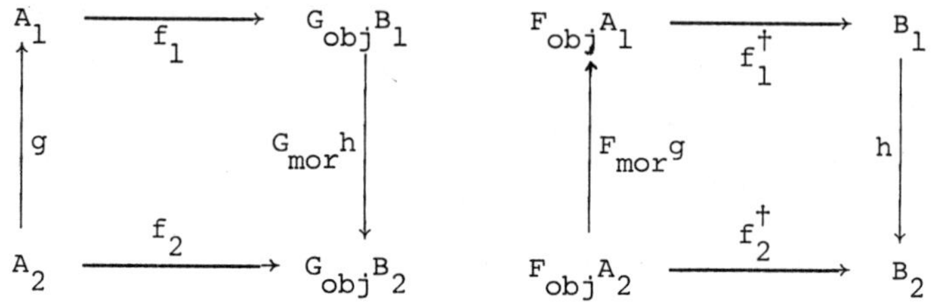

in $\mathcal{C}_1$ in $\mathcal{C}_2$

<u>Theorem</u> (The adjoint functor theorem)

Given a category $\mathcal{C}_1$ whose collection of morphisms between its objects is a set and admitting limits for all diagrams, then a functor $G : \mathcal{C}_1 \longrightarrow \mathcal{C}_2$ has a left adjoint if and only if G preserves limits of all diagrams and satisfies :

For each object B in $\mathcal{C}_2$ there is a set J and morphisms

$$\{f_i : B \longrightarrow GA_i \mid i \in J\}$$

such that every morphism $g : B \longrightarrow GA$ can be written as a composite $g = (Gh) \circ f_i$ for some $i \in J$ and some $h : A_i \longrightarrow A$.

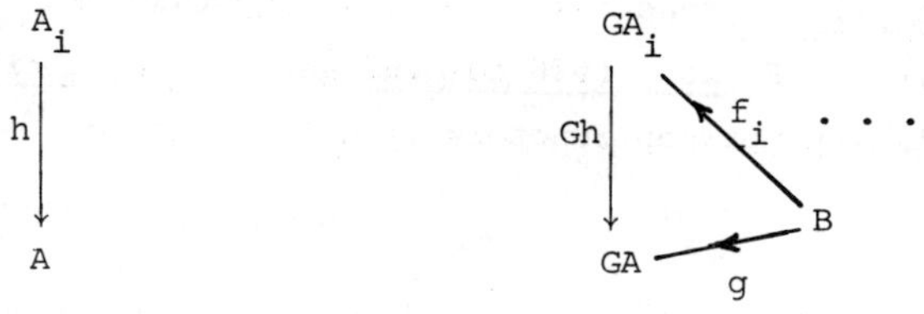

<u>Proof</u>

MacLane [67] p.117. □

Remark

This theorem is important because it suggests the construction process needed for the required left adjoint. We collect below the left adjoints of some common forgetful functors. For constructions and more examples see Herrlich and Strecker [46].

Categories		Adjoint functors	
C_1	C_2	F left adjoint	G right adjoint
Grp	Set	free group	forgetful
Ab	Grp	factor by commutator group	inclusion
Top	Set	discrete space	forgetful
Complete metric spaces	Metric spaces	completion	inclusion
Complete uniform spaces	Uniform spaces	uniform completion	inclusion
Locally convex Top vec	Top vec	locally convex modification	inclusion
Compact Haus	Set	Stone - Čech discrete compactification	forgetful

Example 1

If a functor F has a right (or left) adjoint then that adjoint is uniquely (up to natural equivalence) determined by F. (For a proof see Higgins [47] p.17 et seq.)

Example 2

(Cf. Strooker [97] p.25) Let E be a fixed set and let F be the functor from Set to Set given by the maps

$$F_{obj} : A \longrightarrow A\times E$$

$$F_{mor} : \quad \begin{array}{c} A_1 \\ \downarrow f \\ A_2 \end{array} \quad \longmapsto \quad \begin{array}{c} A_1\times E \\ \downarrow (f, I_E) \\ A_1\times E \end{array}$$

Also let G be the functor from Set to Set given by

$$G_{obj} : A \longrightarrow \{E \longrightarrow A\} = \{\text{all maps from } E \text{ to } A\}$$

$$G_{mor} : \begin{array}{c} A_1 \\ \downarrow f \\ A_2 \end{array} \quad \longmapsto \quad \begin{array}{c} \{E \xrightarrow{h} A_1\} \\ \downarrow \\ \{E \xrightarrow{f\circ h} A_2\} \end{array}$$

Then (F,G) are adjoint functors.

Example 3

(This is the important theorem concerning adjoint functors and limits, cf. Higgins [47] p.52 for proof.)

Let $F : \mathcal{C}_1 \longrightarrow \mathcal{C}_2$ and $G : \mathcal{C}_2 \longrightarrow \mathcal{C}_1$ be functors such that (F,G) is an adjoint pair. Then F preserves right limits and G preserves left limits.

Example 4

The forgetful functor

$$G : \text{Top} \longrightarrow \text{Set}$$

has as left adjoint, the discrete topologising functor

$$F : \text{Set} \longrightarrow \text{Top} : A \longmapsto (A, \text{Sub } A)$$

so (F,G) is an adjoint pair.

Also, G has a right adjoint, the indiscrete topologising functor

$$H : \text{Set} \longrightarrow \text{Top} : A \longmapsto (A, \{A,\emptyset\}),$$

but H has no right adjoint (cf. MacLane [67] p.131).

Example 5

The category Haus of all Hausdorff spaces is a full subcategory (cf. §1.5) of Top, and it is complete.

The inclusion functor

$$G_1 : \text{Haus} \longrightarrow \text{Top}$$

and the forgetful functor

G_2 : Haus ⟶ Set

both have left adjoints, but neither has a right adjoint.
For a proof see MacLane [67] p.131-2.

We conclude this chapter by drawing attention to the extensive bibliography on category theory provided in Herrlich and Strecker [46] pp. 332-380 and also to the exceptionally detailed index in that book. Both will be found very useful by those wishing to pursue further the introductory treatment we have given above.

III Existence of limiting topologies

We can think of topology as the abstract study of continuity for maps between sets. In a sense, having a topology is the least extra structure that a set can be given in order for the notion of continuity to have meaning for maps to and (or) from the set. There are many good introductions to topology : Jameson [54] and Maddox [68] slanted towards normed spaces, Porteous [78] slanted towards topological algebra and Wall [100] slanted towards algebraic topology. As a general reference work Császár [20] is invaluable and we shall use it frequently. The goal of this chapter is to prove the existence of certain limiting topologies by exploiting a partial ordering of topologies on a given set. That will be done in the second section. First we collect for our later convenience some of the basic terminology and elementary constructions. We omit proofs and examples from the first section since they are readily available elsewhere, but mainly we give full details in the second section. There we prove that Top is a complete category and we look at some of its limiting structures. Recent papers on categorical topology can be found in the volume edited by Binz and Herrlich [4].

1. BASIC TOPOLOGY

We shall mainly work from the viewpoint of open sets rather than from neighbourhoods (cf. Jameson [54], Császár [20]). For a more leisurely motivation of the definitions see Dodson and Poston [28] where the goal is to provide for continuity and differentiability on vector spaces as a preliminary to the study of manifolds and applications in relativity. For testing your intuition in topological arguments the study by Steen and Seebach [95] of counterexamples can be helpful!

1.1 Topological space and continuous map

A topological space is a pair (X,T) where X is a set and T is a collection of subsets of X satisfying :

(i) $\emptyset, X \in T$,

(ii) arbitrary unions of sets in T are in T,

(iii) finite intersections of sets in T are in T.

The collection T is called a topology for X.

A continuous map from a topological space (X_1, T_1) to a topological space (X_2, T_2) is a (set-theoretic) map

$$f : X_1 \longrightarrow X_2 \quad \text{such that} \quad f^{\leftarrow}B \in T_1 \quad (\forall B \in T_2).$$

The elements of a topology T for X are called open sets of (X,T), or T-open sets of X, and their complements in X are called closed sets of (X,T), or T-closed sets of X.

Remark

Continuity of maps is preserved by map composition but not by inversion of maps. A continuous bijection with a continuous inverse is called a homeomorphism (i.e. a topological isomorphism) and when one exists between two topological spaces we call them homeomorphic. Being homeomorphic is an equivalence relation $\simeq$ on topological spaces. A surjective map is called open (respectively closed) if the image of every open (closed) set is open (closed). It follows that a bijection is a homeomorphism if and only if it is continuous and open.

Top is the category of all topological spaces with continuous maps among them (cf. II §1.2), we shall write these morphisms in the form

$$f : (X_1, T_1) \longrightarrow (X_2, T_2) .$$

Recall that a pseudo-metric or semimetric on a set X is a map

$$\rho : X \times X \longrightarrow \mathbb{R}$$

satisfying for all $x, y, z \in X$ the properties

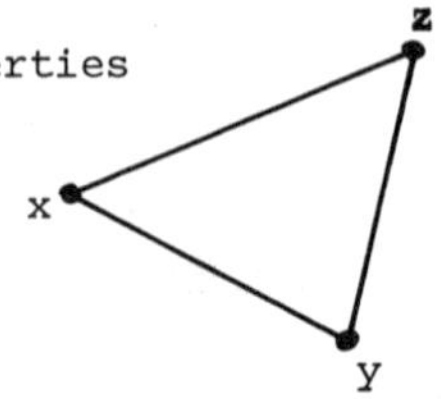

(i) $\rho(x,x) = 0$

(ii) $\rho(x,y) = \rho(y,x)$

(iii) $\rho(x,z) \leqslant \rho(x,y) + \rho(y,z)$.

A set X with such a map is called a pseudo- or semimetric space. We note that by putting $x = y = z$ in (iii) we deduce that ρ is non-negative. If ρ is in fact positive definite, that is to say it satisfies also

(iv) $\rho(x,y) = 0 \Rightarrow x = y$

then we call it a distance function or metric for X. Met is the category of small metric spaces and continuous maps.

This should not be confused with the use of the word 'metric' in physics literature to mean a metric tensor field on a manifold (cf. [28]).

Example

Every semimetric space (X,ρ) is a topological space (X, T_ρ) . Here, any $A \subseteq X$ is a T_ρ -open set if

$$(\forall x \in A)(\exists \text{ real } r>0) : S(x,r) \subset A$$

where

$$S(x,r) = \{y \in X | \rho(x,y) < r\}$$

is the open ball of radius r in (X,ρ).

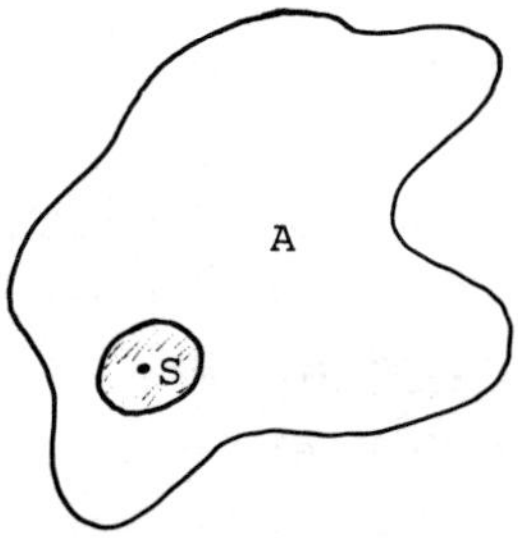

1.2 Neighbourhood, base and sub base

An (open) neighbourhood of any $x \in X$, when (X, T) is a topological space, is any subset of X containing some $A \in T$ with $x \in A$.

A base for a topology T on X is a collection B of subsets of X satisfying

(i) $B \subseteq T$

(ii) every non-empty T-open set is expressible as a union of sets from B .

Example

In any metric space (X,ρ) a base for T_ρ is

$$B = \{S(x,r) \mid x \in X,\ r > 0\}$$

the set of all open balls of positive radius in X.

A sub base for a topology T on X is a collection B_s of subsets of X if the set of all finite intersections of sets from B_s form a base for T. (Recall that the empty set is the smallest finite set, it has no elements.)

Remark

A cover of a set X is a collection C of subsets of X with $X \subseteq \bigcup_{A \in C} A$. Hence any cover can be used as a sub base to generate a topology.

A map f from (X_1, T_1) to (X_2, T_2) is called continuous at $x \in X$ if the inverse image of every neighbourhood of $f(x)$ contains a neighbourhood of x. We say that f is continuous if it is continuous at all points of its domain.

1.3 *Separation axioms*

A topological space (X,T) is called :

(i) T_o if whenever x,y are distinct points in X at least one of them possesses a neighbourhood not containing the other point;

(ii) T_1 if whenever x,y are distinct points in X each of them possesses a neighbourhood not containing the other point;

(iii) T_2 (or Hausdorff) if whenever x,y are distinct points in X they possess neighbourhoods that are mutually disjoint.

It follows that a space is

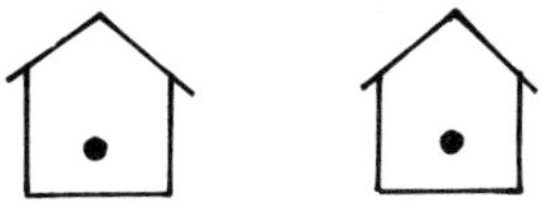

$T_1 \iff$ every point is a closed set

$T_2 \iff$ every point is the intersection of its closed neighbourhoods.

Example

Metric spaces are always T_2, and a semimetric space (X,ρ) is T_o if and only if ρ is a metric.

1.4 *Closure, boundary, compactness, local compactness*

Let (X,T) be a non-empty topological space.

A limit point of a set $A \subseteq X$ is any $p \in X$ such that:

$$p \in N \in T \implies N \setminus \{p\} \cap A \neq \emptyset .$$

The closure A^c of a set $A \subseteq X$ is the union of A and all of its limit points; equivalently A^c is the smallest closed set containing A.

The interior int A of a set $A \subseteq X$ is the largest open set contained in A.

The boundary $\dot{A}$ of a set $A \subseteq X$ is $A^c \setminus \text{int } A$.

The space (X,T) is compact if every open cover contains a finite subcover (cf. §1.2), and locally compact if every point has a compact neighbourhood. We recall that the continuous image of a compact set is compact.

1.5 *Connectedness, denseness, separability*

A topological space (X,T) is called connected if it is not the union of two disjoint non-empty open sets, or equivalently : if the only sets that are both open and closed are $\emptyset$ and X. A component of a topological space is a maximal connected subset. Connectedness is preserved by continuous maps; components are closed sets.

A subset A of the space (X,T) is called dense if $A^c = X$, or equivalently if any neighbourhood of any point of X intersects A . The space (X,T) is called separable if it has a countable dense subset. Since the rationals are countable and dense in the reals it follows that $\mathbb{R}^n$ is separable for each finite n.

The space (X,T) is called

first countable if each point has a countable neighbourhood base

second countable if T has a countable base.

It follows that second countable implies first countable and separable, and Lindelöf's Theorem shows that every open cover of a second countable space has a countable subcovering.

1.6 *Paracompactness and partitions of unity*

In a topological space (X,T) a collection K of subsets of X is called locally finite if every point of X has a neighbourhood which intersects only a finite number of the sets from K. A collection K is called a refinement of a collection L if

$$(\forall A \in K)(\exists B \in L) : A \subseteq B$$

The space (X,T) is called paracompact if every open cover of X has a locally finite refinement which is also an open cover of X.

A non-empty family Φ of continuous real functions on (X,T) is called locally finite if every point of X has a neighbourhood on which all but finitely many of the functions in Φ take the value zero. For such a family, their sum function

$$\sum_{f \in \Phi} f : X \longrightarrow \mathbb{R}$$

is continuous and we call Φ a partition of unity if

$$f \in \Phi,\ x \in X \Longrightarrow f(x) \geq 0,\ \sum_{f \in \Phi} f(x) = 1 .$$

Such a partition of unity is called subordinate to a collection K of subsets of X if

$$(\forall f \in \Phi)(\exists A \in K) : f(x) = 0 \quad \forall x \notin A .$$

It is easily shown that if in (X,T) for every open cover K there is a partition of unity subordinate to K, then (X,T) is paracompact.

Other sufficient conditions for paracompactness are, for example, compactness or pseudo-metrizability. In the case of a (Hausdorff) smooth manifold (cf. Ch. IV), paracompactness is sufficient to ensure the existence of a (smooth) partition of unity which in turn leads to a Riemannian structure and hence metrization.

1.7 *Homotopy, contractibility, covering space*

Given two spaces X, Y and two continuous maps $f_o, f_1 : X \longrightarrow Y$ we say that f_o is homotopic to f_1 if there exists a continuous map (a homotopy)

$$F : X\times[0,1] \longrightarrow Y$$

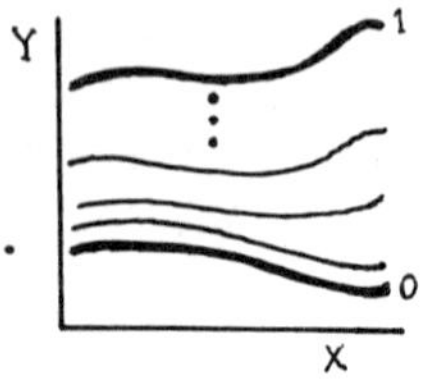

satisfying

$$(\forall x\in X) F(x,0) = f_o(x), \ F(x,1) = f_1(x) .$$

Being homotopic is an equivalence relation $\sim$.

A space X is called <u>contractible</u> if the identity I_X is homotopic to some constant map, $c : X \longrightarrow \{x_o\}$, for some $x_o \in X$.

A space X is called <u>simply connected</u> if it satisfies:

(i) $(\forall x,y\in X) \ \exists$ continuous curve $c : [0,1] \longrightarrow X$ with $c(0) = x, \ c(1) = y$ (<u>arcwise connectedness</u>) ;

(ii) $(\forall x\in X)$ ($\forall$ open neighbourhoods N of x) ($\exists$ open U with $x\in U\subset N$ and U arcwise connected); (<u>local arcwise connectedness</u>) ;

(iii) every closed curve in X is homotopic to a constant curve.

A space $\tilde{X}$ is called a <u>covering space</u> of X if $\tilde{X}$ satisfies (i) and (ii) and there is a continuous surjection $p : \tilde{X} \longrightarrow X$ such that $(\forall x\in X)$ ($\exists$ open neighbourhood N of x) with $p^{\leftarrow}(N)$ a disjoint union of homeomorphs of N . A covering space is called a <u>universal covering space</u> if it is simply connected. (If X is a <u>manifold</u>, cf. IV §1.2, then every covering space has a unique manifold structure making the projection differentiable, and there is a unique <u>universal covering manifold</u>, cf. Steenrod [96] pp. 67-71 and IV §2.3 below. Also for manifolds, connectedness implies arcwise connectedness.)

The <u>fundamental group</u> or <u>first homotopy group</u> $\pi_1(X,x_o)$ of a space X with base point $x_o\in X$ is the group consisting of homotopy equivalence classes of closed curves beginning and ending at x_o . If X is arcwise connected then these groups are isomorphic for all choices of x_o . For introductory treatments of the subject see Singer and Thorpe [90] and Wall [100]. We shall not be venturing into algebraic topology but on a couple of occasions we use a covering space for a manifold. Covering spaces can be made from pairs of points and closed curves through the point, by using homotopy equivalence. The interesting result is that, for each subgroup H of a fundamental group $\pi_1(X)$, a covering space $\tilde{X}_H$

can be found with $\pi_1(\tilde{X}_H) = H$. In particular, if H is trivial then $\tilde{X}_H$ is simply connected and is therefore the universal covering space. Note that Chevalley [12] used the name Poincaré group for what is now usually called the fundamental group; physicists use the name Poincaré group for the Lorentz group plus translations of Minkowski space (cf. DeWitt and DeWitt [23]). A general reference work on algebraic topology is Spanier [92].

2. LIMITING TOPOLOGIES

Given a set X there is a partial order on topologies for X and we shall use various general properties of this relation. Detailed discussions of posets, sets with partial orders, can be found for example in Birkhoff and Bartree [5] several of whose proofs we shall adapt for our purposes.

By space we shall mean topological space.

Since we may simultaneously be considering several topologies on a given set we need to be careful about just calling maps continuous. To emphasize which topologies are involved it is safer to say that a map

$$f : X_1 \longrightarrow X_2$$

from space $(X_1 T_1)$ to (X_2, T_2) is (T_1, T_2) - continuous or to say that

$$f : (X_1 T_1) \longrightarrow (X_2, T_2)$$

is continuous.

We shall use $\square$ to signify the end of a proof or its omission.

Our main result is to establish the completeness of the category Top. Then we look at adjointness properties.

2.1 *Partial order for topologies*

A partial order for a set S is a reflexive, antisymmetric, transitive relation $\leq$ on S.
This means that :

(i) $\leqslant$ is a subset of $S \times S$, the set of all ordered pairs from S (cf. II §2.1 Ex. 3);

(ii) (x,x) is a member of $\leqslant$ for all $x \in S$;

(iii) (x,y) and (y,x) in $\leqslant$ implies $x=y$;

(iv) (x,y) and (y,z) in $\leqslant$ implies (x,z) in $\leqslant$.

The word 'partial' is specifically chosen to emphasize that given a partial order $\leqslant$ on S and $x,y \in S$ then it is possible that neither (x,y) nor (y,x) is in $\leqslant$. When (x,y) is in a partial order $\leqslant$ we say that x and y are comparable, otherwise they are incomparable.

For convenience, we usually abbreviate:

"(x,y) is in $\leqslant$" to "$x \leqslant y$" .

Given a set X we obtain a partial order $\leqslant$ for topologies on X from the partial order $\subseteq$ of subset inclusion by:

$$T_1 \leqslant T_2 \iff (A \in T_1 \implies A \in T_2) .$$

Equivalently

$$T_1 \leqslant T_2 \iff \text{every } T_1\text{-open set in } X \text{ is } T_2\text{-open also.}$$

We shall write $T_1 < T_2$ if $T_1 \neq T_2$ but $T_1 \leqslant T_2$.

Example 1

Let ρ, τ be two semimetrics for X and suppose for some constant $c > 0$, $\rho(x,y) \leqslant c\,\tau(x,y)$, $\forall x,y \in X$. Then $T_\rho \leqslant T_\tau$.

Example 2

$X_{ind} = \{\emptyset, X\}$, the indiscrete topology and $X_{dis} = \text{sub } X$, the discrete topology on a set X are the extreme choices, the smallest and largest respectively.

Remark

Normally when we wish to topologise a set we choose a topology somewhere between these extremes and our choice is usually influenced by which maps to or from other spaces are destined to be continuous.

Example 3

If (Y,T) is any topological space, X is any set and $g : Y \longrightarrow X$, $f : X \longrightarrow Y$ are any maps then we necessarily have:

(i) $g : (Y,T) \longrightarrow (X,X_{ind})$ is continuous,

(ii) $f : (X,X_{dis}) \longrightarrow (Y,T)$ is continuous.

2.2 *Existence of extremal topologies among a finite number*

2.2.1 Theorem

Given any non-empty set $S = \{T_i | i = 1,2,\ldots,n\}$ of a finite number n of topologies for a set X there exist minimal and maximal members of S, T_{min} and T_{max}, given by, respectively:

(i) $(\nexists T_i \in S) : T_i < T_{min}$

(ii) $(\nexists T_i \in S) : T_{max} < T_i$.

Proof

We use the principle of consistent enumeration for partially ordered sets (cf. e.g. [5] p.40).

Let $S_m = \{T_1, T_2, \ldots, T_m\}$.

We construct inductively a permutation β_m of $1,2,\ldots,m$ such that the re-ordering of S_m as

$$T_{\beta_m(1)}, T_{\beta_m(2)}, \ldots, T_{\beta_m(m)}$$

satisfies

$$T_{\beta_m(i)} < T_{\beta_m(j)} \implies i < j .$$

Necessarily, β_1 is the identity .

Suppose we have β_{r-1} for some $r \geqslant 1$ and let k be the first integer such that

$$T_1 < T_{\beta_{r-1}(i)} = T_{\beta_{r-1}(k)} .$$

Hence we define the permutation

$$\beta_r(i) = \begin{cases} i & \text{if } i < k \\ r & \text{if } i = k \\ i-1 & \text{if } i > k \end{cases}$$

which puts T_r between $T_{\beta_{r-1}(k-1)}$ and $T_{\beta_{r-1}(k)}$.

So if $\{T_{\beta_r(i)}, T_{\beta_r(j)}\} \subset S_{r-1}$ then

$$T_{\beta_r(i)} < T_{\beta_r(j)} \Longrightarrow i < j .$$

Also,

$$T_r = T_{\beta_r(k)} < T_{\beta_r(j)} \Longrightarrow k < j ;$$

and

$$T_{\beta_r(i)} < T_{\beta_r(k)} = T_r \Longrightarrow i < k$$

because $T_{\beta_r(k)} < T_{\beta_r(k+1)}$ by construction and so

$T_{\beta_r(i)} < T_{\beta_r(k+1)}$ by transitivity of $\leqslant$ and then by construction $i < k + 1$, but since β_r is a bijection we know that $i \neq k$.

Finally, we apply β_n to the whole set $S = S_n$ and the required minimal and maximal elements are

$$T_{min} = T_{\beta_n(1)},\ T_{max} = T_{\beta_n(n)} . \qquad \square$$

Remark

The generalisation of this that we want for arbitrary sets is like that needed in the real number system, and the same terminology is used.

2.3 *Uniqueness of sup and inf topologies*

Given any set S of topologies for a set X we say that :

(i) a topology T_ℓ for X is a lower bound of S if $T_\ell \leqslant T$ for all $T \in S$;

(ii) a topology T_u for X is an <u>upper bound</u> of S if $T \leqslant T_u$ for all $T \in S$;

(iii) a topology $T_[$ for X is a <u>greatest lower bound</u> of S, written $T_[= \inf S$, if $T_[$ satisfies (i) and also if T_ℓ is any other lower bound of S then $T_\ell \leqslant T_[$;

(iv) a topology $T_]$ for X is a <u>least upper bound</u> of S, written $T_] = \sup S$, if $T_]$ satisfies (ii) and also if T_u is any other upper bound of S then $T_] \leqslant T_u$.

Our terminology in (iii) and (iv) implies uniqueness, which we prove now.

2.3.1 <u>Theorem</u>

For any non-empty set S of topologies for a set X , if they exist then inf S and sup S are unique.

<u>Proof</u>

We prove the case for inf S, that for sup S is similarly patterned around the antisymmetry of $\leqslant$.

Suppose that $T_[$ and $T_($ are both greatest lower bounds for S . Then $T_[\leqslant T_($ and $T_(\leqslant T_[$. Hence by property §2.1 iii, $T_[= T_($. □

As was to be expected, inf S and sup S do exist; our next result shows how to construct them.

2.4 *Existence of sup and inf topologies*

2.4.1 <u>Theorem</u>

Given $S = \{T_x | x \in A\}$ an arbitrary set of topologies for a set X , there exist unique greatest lower and least upper bounding topologies

$$T_[= \inf S$$

$$T_] = \sup S .$$

Proof

If $S = \emptyset$ then we have (cf. §2.1 Ex. 1) $T_{[} = X_{ind}$ and $T_{]} = X_{dis}$, unique by §2.3 .

For $S \neq \emptyset$ we need only prove existence because §2.3 guarantees uniqueness. We begin with

$$T_{[} = \{G \in \text{sub } X \mid \forall \alpha \in A,\ G \in T_\alpha\} ,$$

the intersection of topologies in the set S, that is by using the open sets common to all members of S. We need to check the axioms §1.1 (i) - (iii). Evidently they hold because each T_α is itself a topology. Hence $T_{[}$ is a topology and plainly a lower bound of S. Suppose that T_ℓ is another lower bound of S. Then we argue:

$$T_\ell \leq T_\alpha \qquad (\forall \alpha \in A) \qquad \text{(by §2.3 i)}$$

$$G \in T_\ell \Rightarrow G \in T_\alpha \qquad (\forall \alpha \in A) \qquad \text{(by §2.1)}$$

$$\Rightarrow G \in T_{[} , \quad \text{so } T_\ell \leq T_{[} .$$

Next we construct sup S by using as sub base

$$B_{]} = \{G \in \text{sub } X \mid \alpha \in A : G \in T_\alpha\} ;$$

so $T_{]}$ is the topology generated (cf. §1.2) by arbitrary unions and finite intersections of the members of $B_{]}$. (Of course, the union of topologies in S might not be a topology.) We see that $T_{]}$ is indeed an upper bound of S by :

$$G \in T_\alpha \Rightarrow G \in B_{]} \Rightarrow G \in T_{]}$$

$$\Rightarrow T_\alpha \leq T_{]} \qquad (\forall \alpha \in A) .$$

Suppose that T_u is another upper bound. Then we have

$$G \in T_{]} \Rightarrow \begin{cases} \text{either } G = B_1 \cap B_2 & \text{for some } B_i \in B \\ \text{or} \quad G = \bigcup_{\lambda \in J} B_\lambda & \text{for some } B_\lambda \in B,\ \lambda \in J . \end{cases}$$

Hence, either $(\exists \alpha_1, \alpha_2 \in A) : B_i \in T_{\alpha_i}, \ i = 1,2,$

$$\text{or } (\exists \alpha_\lambda \in A, \lambda \in J) : B_\lambda \in T_{\alpha_\lambda} \ \forall \lambda \in J .$$

In either case we have $G \in T_u$, so $T_u \leq T_{]}$. □

Remark

We obtain two corollaries that help us see how $T_{\rceil}$ is constructed.

Corollary 1

If for each $\alpha\in A$, B_α is a sub base for T_α then $B = \bigcup_{\alpha\in A} B_\alpha$ is a sub base for $T_{\rceil} = \sup\{T_\alpha \mid \alpha\in A\}$.

Proof

By the theorem we observe that the members of B are in $T_{\rceil}$. Suppose that $G\in T_{\rceil}$; we must construct it as a finite intersection or union of sets from B . If $G = \emptyset$ we can intersect the empty set of sets from B . Otherwise we know by construction of $T_{\rceil}$ in the theorem that

$$G = \bigcup_{r\in D} G_r \qquad \text{for some indexing set } D$$

with

$$G_r = \bigcap_{i=1}^{r} H_i \qquad \text{for some } H_i \in B_{\rceil} .$$

Now fix $r\in D$, then for each $i = 1,2,\ldots,n_r$ there is $\alpha_i\in A$ with $H_i\in T_{\alpha_i}$. So each H_i is a union of finite intersections from the sub base B_{α_i} , hence similarly constructible from B . It follows that G is a union of finite intersections of sets from B . □

Corollary 2

Let $x\in X$ and for each $\alpha\in A$ suppose that $\mathcal{N}_\alpha(x)$ is a base for T_α-neighbourhoods of x ; this means that (cf. §1.2) every open set containing x contains a set constructible from finite intersections of sets from $\mathcal{N}_\alpha(x)$. Then finite intersections from the set

$$\{N_\alpha(x) \in \mathcal{N}_\alpha(x) \mid \alpha\in A\}$$

constitute a base for $T_{\rceil}$-neighbourhoods of x .

Proof

The proposed base sets are of the form

$$\bigcap_{i=1}^{n} N_{\alpha_i}(x) \,, \quad \alpha_i \in A, \quad N_{\alpha_i}(x) \in \mathcal{N}_{\alpha_i}(x) \,.$$

Suppose $x \in G_{\alpha_i} \in T_{x_i}$ with $G_{\alpha_i} \subset N_{\alpha_i}(x)$, $i = 1,2,\ldots,n$.

Then $x \in \bigcap_{i=1}^{n} G_{\alpha_i} \subset \bigcap_{i=1}^{n} N_{\alpha_i}(x)$

and

$$\bigcap_{i=1}^{n} G_{\alpha_i} \in T_{]} \text{ by construction,}$$

so the proposed sets are indeed $T_{]}$ - neighbourhoods of x .

Given some $T_{]}$ - neighbourhood $N(x)$ of x we can find from the sub base $B_{]}$ some $G_{\alpha_i} \in T_{\alpha_i}$, $i = 1,2,\ldots,n$ for some n and

$$x \in \bigcap_{i=1}^{n} G_{\alpha_i} \subset N(x) \,.$$

But then $x \in G_{\alpha_i}$ for $i = 1,2,\ldots,n$ and so each G_{α_i} is a finite intersection of sets from $\mathcal{N}_{\alpha_i}(x)$. Therefore, $\bigcap_{i=1}^{n} G_{\alpha_i}$ is a finite intersection from

$$\{N_\alpha(x) \in \mathcal{N}_\alpha(x) \mid \alpha \in A\}$$

So we have found a $T_{]}$ - neighbourhood of x contained in $N(x)$ and constructed in the form $\bigcap_{i=1}^{n} N_{\alpha_i}(x)$. □

2.5 *Coinduced and induced topologies*

We can construct a unique topology on a set X if we have a map from the set to a topological space or from a topological space to the set. In each case we take the appropriate limiting topology that makes the map continuous.

Lemma

Let (X,T) be a topological space and for some set Y suppose that we have

either (i) a map $f : Y \longrightarrow X$

or (ii) a map $g : X \longrightarrow Y$.

Then, in case (i)

$$\overleftarrow{f}T = \{\overleftarrow{f}G \mid G \in T\}$$

is the smallest topology for Y that makes f continuous, and in case (ii)

$$gT = \{H \subseteq Y \mid \overleftarrow{g}H \in T\}$$

is the largest topology for Y that makes g continuous.

Proof

In each case the topology axioms are satisfied because of the way that maps act on intersections and unions. If either Y or X is empty then it has the unique topology $\{\emptyset\}$ and f or g is the empty map. We prove the extremal properties in turn:

(i) Suppose that T_ℓ is a topology for Y and

$$f : (Y,T_\ell) \longrightarrow (X,T) \qquad *$$

is continuous. By construction, any open set of $(X,\overleftarrow{f}T)$ is of the form $\overleftarrow{f}G$ for some $G \in T$. Then, by hypothesis *, $\overleftarrow{f}G \in T_\ell$ and so

$$\overleftarrow{f}T \leqslant T_\ell \ .$$

(ii) Suppose that T_u is a topology for Y with

$$g : (X,T) \longrightarrow (Y,T_u)$$

continuous. Let $H \in T_u$. Then $\overleftarrow{g}H \in T$ by hypothesis and so $H \in gT$ by construction and hence

$$T_u \leqslant gT \ . \qquad \square$$

Example

The subspace topology on $S \subseteq X$ for a space (X,T) is just the restriction of T to S ,

$$T|_S = \{G \cap S \mid G \in T\} \ .$$

This coincides with $\overleftarrow{f}T$ where f is the inclusion map of S in X. Observe that f need not be open: a closed interval of R is open in the subspace topology of that interval.

Remark

For the circumstances of the Lemma we call:

(i) $\overleftarrow{f}T$ the coinduced (or inverse image) topology by f from (X,T)

(ii) gT the induced (or quotient) topology by g from (X,T).

Corollary 1

We collect some properties of the above topologies, for proofs see Császár [20].

(i) If B is a base (sub base) for T then $\overleftarrow{f}B$ is a base (sub base) for $\overleftarrow{f}T$.

(ii) If T' is another topology for X with $T' \leqslant T$ then $\overleftarrow{f}T' \leqslant \overleftarrow{f}T$.

(iii) If $\{T_\alpha | \alpha \in A\}$ are topologies for X then $\{\overleftarrow{f}T_\alpha | \alpha \in A\} = \overleftarrow{f} \sup \{T_\alpha | \alpha \in A\}$,

(iv) A set F is closed in the topology gT

$$\iff \overleftarrow{g}F \text{ is closed in } T.$$

(v) A map

$$h : (X_1, T) \longrightarrow (X_2, T_2)$$

is continuous

$$\iff \overleftarrow{h}T_1 \leqslant T_1$$

$$\iff T_2 \leqslant hT_1 .$$

(vi) If a map

$$h : (X_1, T_1) \longrightarrow (X_2, T_2)$$

is continuous and open
(or continuous and closed)

then $hT_1 = T_2$ □

2.5.1 Theorem

Let $\{(X_\alpha, T_\alpha) \mid \alpha \in A\}$ be a collection of topological spaces and suppose that Y is a set.

(i) If for each $\alpha \in A$ there is a map $f_\alpha : Y \longrightarrow X_\alpha$, then

$$\sup \{\overleftarrow{f_\alpha} T_\alpha \mid \alpha \in A\}$$

is the <u>smallest</u> topology for which every f_α is continuous.

(ii) If for each $\alpha \in A$ there is a map $g_\alpha : X_\alpha \longrightarrow Y$, then

$$\inf\{g_\alpha T_\alpha \mid \alpha \in A\}$$

is the <u>largest</u> topology for which every g_α is continuous.

Proof

In each case we know by §2.4 that the candidate exists, is unique and a topology because our Lemma showed that the constituent sets are topologies. By corollary 1(v) we see that f_α is continuous if and only if

$$\overleftarrow{f_\alpha} T_\alpha \leqslant \sup \{\overleftarrow{f_\alpha} T_\alpha \mid \alpha \in A\}$$

and so we do have the smallest topology in case (i). Similarly, in case (ii), continuity of g_α is equivalent to

$$\inf \{g_\alpha T_\alpha \mid \alpha \in A\} \leqslant g_\alpha T_\alpha$$

so we have the largest topology here. □

Corollary 2

(i) If, for each $\alpha \in A$, B_α is a sub base for T_α then

$$\{\overleftarrow{f_\alpha} G \mid G \in B_\alpha, \alpha \in A\}$$

is a sub base for

$$\sup \{\overleftarrow{f_\alpha} T_\alpha \mid \alpha \in A\} .$$

(ii) If for each $\alpha \in A$ there is another topology T'_α for X_α with $T_\alpha \leqslant T'_\alpha$ then $\sup \{\overleftarrow{f_\alpha} T_\alpha \mid \alpha \in A\} \leqslant \sup \{\overleftarrow{f_\alpha} T'_\alpha \mid \alpha \in A\}$ and $\inf \{g_\alpha T_\alpha \mid \alpha \in A\} \leqslant \inf \{g_\alpha T'_\alpha \mid \alpha \in A\}$.

Proof

Császár [20] . □

2.6 *Product and coproduct topologies*

From the definition of product in II §2.2 and its existence in the category Set, we know that for any collection $\{(X_i,T_i)\,|\,i\in J\}$ of topological spaces there exists in Set their product set $\prod_{i\in J} X_i$. Plainly, the 'right' topology to put on this set is the smallest one that makes every projection map (existing in Set) $\pi_j : \prod_{i\in J} X_i \longrightarrow X_j$ continuous.

Hence, we define the product topology of the T_i by

$$\prod_{i\in J} T_i = \sup \{\overleftarrow{\pi_i} T_i\,|\,i\in J\} .$$

2.6.1 Theorem

The product space $\left(\prod_{i\in J} X_i, \prod_{i\in J} T_i\right)$ is the product in Top of the objects $\{(X_i,T_i)\,|\,i\in J\}$.

Proof

Existence is assured by our previous work. Also, $\prod_{i\in J} X_i$ is unique up to isomorphism in Set,being a left limit (cf. II §2.1) of $\{X_i\,|\,i\in J\}$, and $\prod_{i\in J} T_i$ is unique because by §2.3 sup gives a unique topology.

We need to show that given any continuous

$g_i : (K,T) \longrightarrow (X_i,T_i)$, $\forall i\in J$

there is a unique continuous

$\ell : (K,T) \longrightarrow \left(\prod_{i\in J} X_i, \prod_{i\in J} T_i\right)$

with $g_i = \pi_i \circ \ell \quad \forall i\in J$.

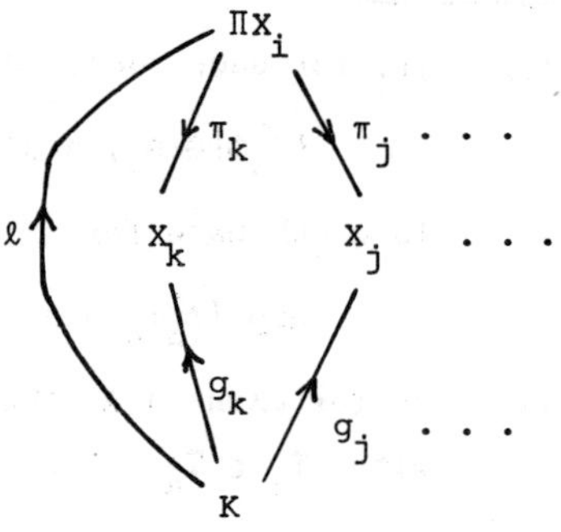

Our candidate for ℓ is obviously the one from Set, unique as a map,

$$\ell : K \longrightarrow \prod_{i \in J} X_i \; : \; x \longmapsto (g_i(x))_{i \in J} \; .$$

It remains to show continuity.

Let G be open in $\prod_{i \in J} X_i$.

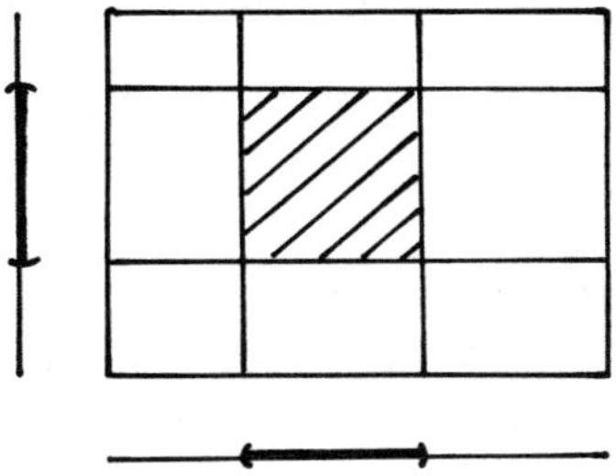

Then by §2.4 either

(i) $\quad G = \bigcap_{k=1}^{n} \pi_{i_k}^{\leftarrow} G_{i_k} \, , \; G_{i_k} \in T_{i_k}$

or

(ii) $\quad G = \bigcup_{i \in J_o} \pi_i^{\leftarrow} G_i \, , \; G_i \in T_i \, , \; J_o \subseteq J$.

In case (i) $\quad \ell^{\leftarrow} G = \bigcap_{k=1}^{n} \ell^{\leftarrow} \circ \pi_{i_k}^{\leftarrow} G_{i_k}$

$$= \bigcap_{k=1}^{n} g_{i_k}^{\leftarrow} G_{i_k}$$

which is open in K since each g_{i_k} is continuous.

In case (ii) $\quad \ell^{\leftarrow} G = \bigcup_{i \in J_o} \ell^{\leftarrow} \circ \pi_i^{\leftarrow} G_i$

$$= \bigcup_{i \in J_o} g_i^{\leftarrow} G_i$$

which is also open by continuity of the g_i . Hence the given map ℓ is continuous. □

In a dual manner we define the <u>coproduct topology</u> $\coprod_{i \in J} T_i$ on a family of spaces $\{(X_i, T_i) \mid i \in J\}$ to be the <u>largest</u> one that makes continuous all of the set-theoretic injections (cf. II §2.2)

$$f_j : X_j \longrightarrow \coprod_{i \in J} \; .$$

That is to say,

$$\coprod_{i \in J} T_i = \inf \{ f_j T_j \mid j \in J \} \; .$$

2.6.2 <u>Theorem</u>

The <u>coproduct space</u> or <u>topological disjoint sum</u> $\left(\coprod_{i\in J} X_i, \coprod_{i\in J} T_i\right)$ is the coproduct in Top of the objects $\{(X_i, T_i) \mid i\in J\}$.

<u>Proof</u>

Existence and uniqueness in Top follow as for the previous theorem. We must show that our candidate has the right limit universal property (cf. II §2.1) : given any continuous maps

$$g_i : (X_i, T_i) \longrightarrow (K,T), \ \forall i\in J ,$$

there exists a unique continuous map

$$r : \left(\coprod_{i\in J} X_i, \coprod_{i\in J} T_i\right) \longrightarrow (K,T)$$

such that

$$g_i = r \circ f_i \quad \forall i\in J .$$

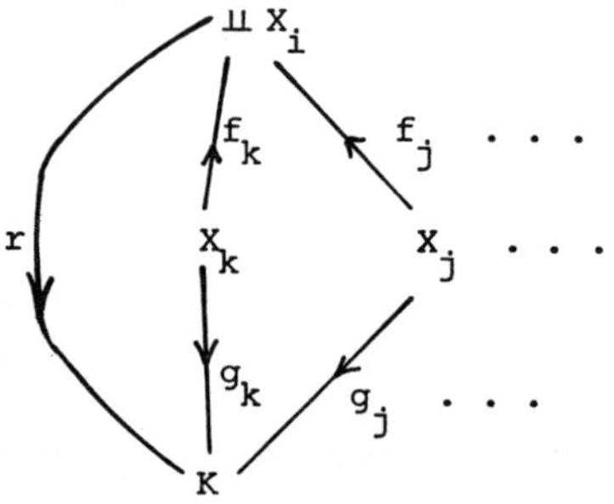

From Ex. 2.1.4 we already have a candidate for r, it is the unique map

$$r : f_i(x) \longmapsto g_i(x) \quad \text{if} \quad x\in X_i \overset{f_i}{\hookrightarrow} \coprod_{i\in J} X_i .$$

We need only show continuity.

Suppose then that G is open in K; we must show that $\overleftarrow{r}G$ is open in $\coprod_{i\in J} X_i$.

Since each f_i is continuous by construction, we have:

$$\overleftarrow{f_i}G = G_i \in T_i \quad \forall i\in J ,$$

and by construction of r

$$\overleftarrow{f_i} \circ \overleftarrow{r}G = \overleftarrow{g_i}G .$$

So, as required, we have shown that, for each $i\in J$, $\overleftarrow{f_i}(\overleftarrow{r}G)$ is open in X_i because each g_i is continuous by hypothesis. □

<u>Remark</u>

It is of course important to know which properties of constituent spaces persist in the product and coproduct spaces. We collect a few of these properties, for proofs and more results see e.g.

Császár [20] and for counterexamples Steen and Seebach [95].

Property of all factor spaces (X_i, T_i) $i \in J$	Whether preserved by product and coproduct					
	J finite		J countable		J uncountable	
	Π	$\amalg$	Π	$\amalg$	Π	$\amalg$
Compact	Yes	Yes	Yes	No	Yes	No
Locally compact	Yes	Yes	No	Yes	No	Yes
First countable	Yes	Yes	Yes	Yes	No	Yes
Second countable	Yes	Yes	Yes	Yes	No	No
Separable	Yes	Yes	Yes	Yes	No	No
Paracompact	No	Yes	No	Yes	No	Yes
T_o, T_1, T_2 = Hausdorff	Yes	Yes	Yes	Yes	Yes	Yes
Connected	Yes	No	Yes	No	Yes	No

2.7 Completeness of Top

We know from Ch. II §2.5 Ex. 9 (cf. Higgins [47] p.54) that a category is left complete if it admits products and equalizers and right complete if it admits coproducts and coequalizers. We have established products and coproducts in Top, in the previous section. Also, in Ch. II §2.4 Exs. 1,2 we constructed equalizers and coequalizers in Set, next we provide them with their natural topologies.

2.7.1 Theorem

Top is a complete category.

Proof

After §2.6 we need only establish equalizers and coequalizers. So consider in Top the diagram

$$\Delta : (X_1, T_1) \underset{a_2}{\overset{a_1}{\rightrightarrows}} (X_2, T_2)$$

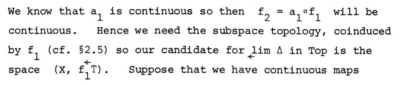

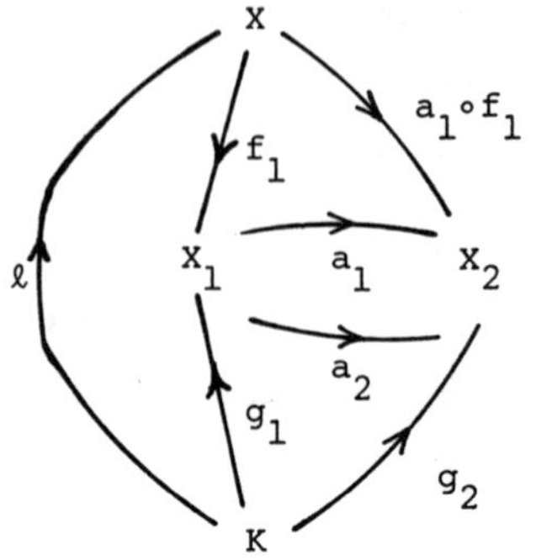

(i) The equalizer set is (cf. II §2.4)

$X = \{x \in X_1 | a_1(x) = a_2(x)\}$

and we have to provide it with a topology that makes continuous the inclusion

$$f_1 : X \hookrightarrow X_1 ,$$

which is an injective set map.
We know that a_1 is continuous so then $f_2 = a_1 \circ f_1$ will be continuous. Hence we need the subspace topology, coinduced by f_1 (cf. §2.5) so our candidate for $\underleftarrow{\lim}\, \Delta$ in Top is the space $(X, \overleftarrow{f_1}T)$. Suppose that we have continuous maps

$$(K, T_o) \xrightarrow{g_1} (X_1, T_1)$$
$$(K, T_o) \xrightarrow{g_2} (X_2, T_2)$$

commuting with the a_1, a_2. We have the map:

$$\ell : K \longrightarrow X : y \longmapsto y'$$

where $\{y'\} = \overleftarrow{f_1} g_1(y)$

because by commutativity a_1 and a_2 agree on X in the image of g_1. We need to establish continuity of ℓ. Let G be open in $(X, \overleftarrow{f_1}T_1)$. Then we have some $S \in T_1$ with $G = \overleftarrow{f_1}S$ by Lemma 2.5. Thus we deduce

$$\overleftarrow{\ell}G = \overleftarrow{\ell} \circ \overleftarrow{f_1}S = \overleftarrow{g_1}S \quad (\text{since } f_1 \circ \ell = g_1)$$

which is open in K since g_1 is continuous.

(ii) The coequalizer set is (cf. II §2.4)

$$Y = X_2/\sim$$

where $\sim$ is the smallest equivalence relation on X_2 that contains

$$\{(a_1(x),\ a_2(x))\,|\,x\in X_1\}\ .$$

Also,

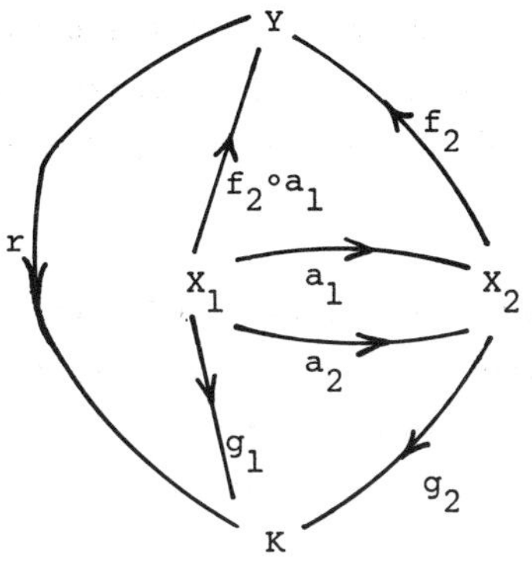

$$f_2 : X_2 \longrightarrow Y : x \longmapsto [x]_\sim$$

is the surjection onto classes and

$$f_1 = f_2 \circ a_1\ .$$

The right topology is that induced by f_2 (cf. §2.5) so our candidate right limit in Top is $(Y, f_2 T_2)$.

Suppose that we are given continuous maps

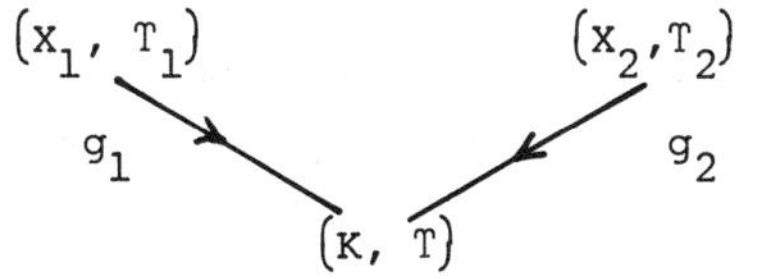

commuting with the a_1, a_2. Then we already have in Set the unique map

$$r : Y \longrightarrow K : [x]_\sim \longmapsto g_2(x)$$

satisfying $r \circ f_2 = g_2$, $r \circ f_1 = r \circ (f_2 \circ a_1) = g_1$.

This map is well-defined for suppose that

$$y \in [x]_\sim$$

then

$$a_1(x) = a_2(y) \quad \text{and so} \quad g_2 \circ a_1(x) = g_2 \circ a_2(y)$$

or $\quad a_1(x) = a_1(y) \quad$ so $\quad g_2 \circ a_1(x) = g_2 \circ a_1(y)$

or $\quad a_2(x) = a_2(y) \quad$ so $\quad g_2 \circ a_2(x) = g_2 \circ a_2(y)$

hence $r([x]_\sim) = r([y]_\sim)$ whenever $(x \sim y)$.

It remains to show that r is continuous. Let G be open in (K,T) . Now, by §2.5, H is open in $(Y, \overleftarrow{f_2} T_2)$ only if $\overleftarrow{f_2} H$

is open in T_2 . But

$$\overleftarrow{f_2}(\overleftarrow{r}G) = \overleftarrow{(r \circ f_2)}G = \overleftarrow{g_2}G$$

and so $\overleftarrow{r}G$ is indeed open in $(Y, \overleftarrow{f_2}T_2)$.

Therefore, Top admits coequalizers and by (i) equalizers; since by §2.6 it also admits products and coproducts Top is complete : it has left and right limits for all of its diagrams. □

Remark

We have previously noted that if a left (or right) limit exists in a category $\mathcal{C}$ then it is unique up to isomorphism in $\mathcal{C}$. In Top, an isomorphism is a homeomorphism, a bijective continuous map with continuous inverse.

2.8 *Projective limit and inductive limit topologies*

Products and coproducts arise from collections of isolated spaces. In general we are interested in collections of spaces with some maps among them; then we can use products and coproducts as the basis for constructing limits.

Our concern here is with limits of diagrams in Top like

$$\Delta = \{p_{ij} : (X_i,T_i) \longrightarrow (X_j,T_j) \mid i\leqslant j,\ i,\ j\in J\}$$

where (i) J is some non-empty set of indices with partial order $\leqslant$

(ii) $p_{ii} = I_{X_i}$ $(\forall i\in J)$

(iii) $p_{ik} = p_{jk} \circ p_{ij}$ whenever $i\leqslant j\leqslant k$.

The projective limit space is the left limit of Δ and the inductive limit space is the right limit of Δ . These spaces exist by the completeness of Top (cf. §2.7) and, like all limits there, are unique up to homeomorphism. Further discussion will be found in Kowalsky [59]. As we pointed out in Ch. II §2.5 Ex. 8, set products and coproducts lie at the basis of limits in Top and we shall prove this next for the present cases. We shall see in Chapter V how the projective limit is used to construct a boundary for spacetime (cf. [15], [26]).

2.8.1 Theorem

With Δ as above, the left limit object is given by the subspace (cf. §2.6) of the product space

$$X = \{(x_i)_{i\in J} \in \prod_{i\in J} X_i \mid j\leq k\in J \Rightarrow p_{jk}(x_j) = x_k\}$$

with associated morphisms the restrictions of the projections (cf. §2.6 and II §2.2)

$$p_k = \pi_k|_X : (x_i)_{i\in J} \longmapsto x_k .$$

Proof

Existence is assured by completeness of Top (cf. §2.7) and so we only need to establish the left limit property. Suppose then that we have continuous maps

$$g_i : (K,T) \longrightarrow (X_i,T_i)$$

for all $i\in J$, commuting with the contiguous p_{jk}.

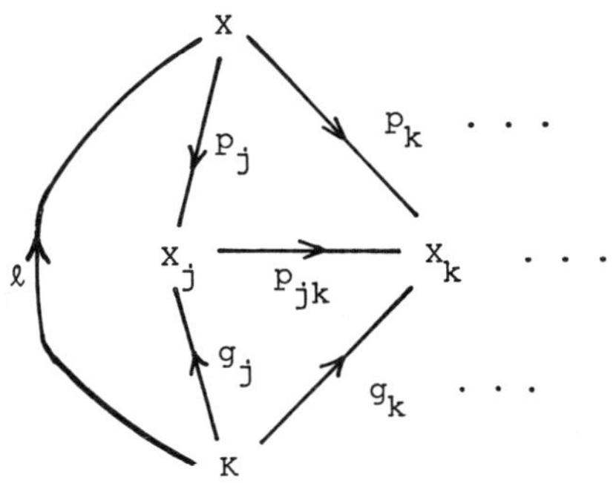

From §2.6 we have a continuous map

$$\ell : K \longrightarrow \prod_{i\in J} X_i : x \longmapsto (g_i(x))_{i\in J} .$$

We can see that for the given g_i it takes values in X. For if $j\leq k$ then we deduce

$$(\forall x\in K)\ p_{jk}\circ g_j(x) = g_k(x) ,$$

by commutativity. □

2.8.2 Theorem

With Δ as above, the right limit object is given by the set of equivalence classes

$$Y = \{[x]_\sim \mid x \in \coprod_{i\in J} X_i \}$$

where $\sim$ is the relation

$$x \sim y \iff \exists i \leq k,\ j \leq k \quad \text{with} \quad p_{ik}(x) = p_{jk}(y)$$

and associated morphisms are the projections onto the classes (cf. II §2.2) :

$$f_i : X_i \longrightarrow Y : x \longmapsto [x]_\sim$$

and the topology is that induced by projection onto classes from the space

$$\left(\coprod_{i \in J} X_i \ , \ \coprod_{i \in J} T_i \right) .$$

Proof

We know by §2.7 that a right limit exists. We show that our candidate meets the requirements, then we know that it is unique up to homeomorphism.

Suppose that we have continuous maps

$$g_i : (X_i, T_i) \longrightarrow (K,T)$$

for all $i \in J$, commuting with the contiguous p_{jk}. From §2.6 we have a continuous map

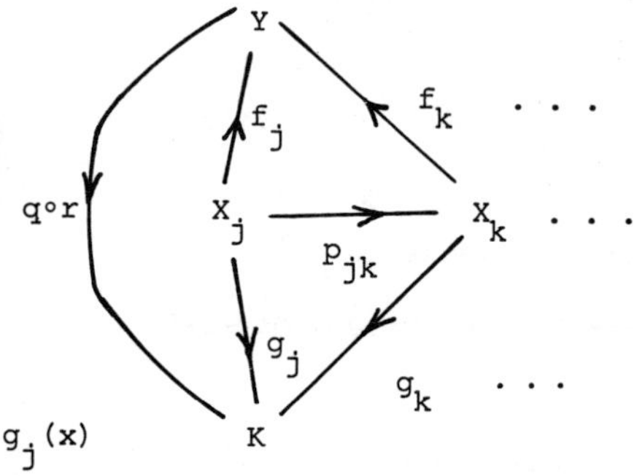

$$r : \coprod_{i \in J} X_i \longrightarrow K : f_j(x) \longmapsto g_j(x)$$

when $x \in X_j$.

Next we note that $\sim$ is an equivalence relation. The proposed topology on Y is induced by

$$q : \coprod_{i \in J} X_i \longrightarrow Y : x \longmapsto [x]_\sim ,$$

which is to say that

$$q : \left(\coprod_{i \in J} X_i , \coprod_{i \in J} T_i \right) \longrightarrow \left(Y, q \coprod_{i \in J} T_i \right)$$

is continuous. It follows that the composite

$$q \circ r : \left(Y, q \coprod_{i \in J} T_i \right) \longrightarrow (K,T)$$

is also continuous and satisfies

$$g_j = q \circ r \circ f_j \quad \forall j \in J .$$

If also there is a continuous map

$$s : \left(Y,\ q \coprod_{i\in J} T_i\right) \longrightarrow (K,T)$$

satisfying

$$g_j = s \circ f_j \quad \forall j \in J ,$$

then

$$(\forall y \in Y)(\exists j \in J,\ x \in X_j) : f_j(x) = [x]_\sim = y .$$

Hence

$$g_j(x) = s[x]_\sim = s(y)$$

but then

$$q \circ r(y) = q \circ r(f_j(x) = g_j(x) = s(y) .$$

So the map $q \circ r$ is unique. □

Example 1

(Cf. Császár [20] p.297, Ex. 2)

Let (X_o, T_o) be a topological space, let $J = N = \{1,2,\ldots\}$ with the usual ordering, and suppose we have subsets $\{X_i | i \in N\}$ of X_o giving the inclusion diagram

$$\emptyset \neq X_1 \underset{p_{12}}{\hookrightarrow} X_2 \underset{p_{23}}{\hookrightarrow} X_3 \ldots X_i \underset{p_{ii+1}}{\hookrightarrow} X_{i+1} \ldots$$

for subspaces $\{(X_i, T_i) | i \in N\}$, so $T_i = T_o|_{X_i}$.

Then the projective limit space is homeomorphic to the subspace $\bigcap_{i=1}^{\infty} X_i$ of X_o

Example 2

(Cf. Császár [20] p. 330, Ex. 18)

Let (X_o, T_o) be a topological space and $\{X_i | i \in J\}$ any collection of subsets of X_o, each X_i having the subspace topology T_i. Then (X_o, T_o) is homeomorphic to the inductive limit of the inclusion diagram.

IV Manifolds and bundles

We can think of a differentiable structure for a topological space as the provision for analysis on that space. Quite generally, a derivative of a map f is a linear map (a morphism from Vec) that locally is an approximation to f itself. This is a natural abstraction from real analysis and the least extra structure needed to support it is a consistent establishment of local tangent vector spaces. A differentiable manifold is a topological space on which such derivatives can be defined. An introductory treatment of manifolds can be found in Dodson and Poston [28], which begins with vector spaces and proceeds to a rigorous formulation of relativistic spacetime; mainly our notation will follow this text. Manifolds and differentiable maps form a category, Man, but it is not complete because of the absence of certain pullbacks, quotients and coproducts (cf. Lang [60] and Hirsch [49]).

Some superstructure is inevitably present on a differentiable manifold, namely its tangent bundle, which is a differentiable manifold made up from the local tangent vectors. In applications, other such vector bundles are useful, and indeed so are the more general fibre bundles which do not break down into local vector spaces but into local differentiable manifolds. Thus, all scalar, vector and tensor fields in physics are sections (cf. II §1.10, Ex. 4) of appropriate vector bundles over a spacetime manifold. On the other hand, a parallelization (a smooth choice of basis for tangent vector spaces over a manifold, cf. §2.4 below) is a section of the fibre bundle of linear frames (ordered bases for tangent spaces).

In this chapter we shall collect some basic terminology for manifolds and bundles and provide for certain constructions that we shall need for spacetime geometry in the next chapter.

There are many excellent texts on manifolds and bundles so we shall not attempt more than an introductory guide to the subject. Basic manifold theory will be found in an eminently digestible form in Spivak [94], Brickell and Clark [9], Bishop and Crittenden [6], and more advanced material is given in Kobayashi and Nomizu [57, 58], Lang [60], Schwartz [87] and Hirsch [49]. Several of these books make some use of bundles but for a detailed treatment of bundle theory see Husemoller [52] and of course the classic Steenrod [96]. Intimately connected with the bundle geometry constructed over spacetime manifolds is the notion of a Lie group, being a group that is also a manifold. A detailed account of the theory of these entities is given by Hochschild [50]. Applications of bundle theory to the analysis of spacetime singularities are studied in Dodson [26], to which we shall refer in the sequel.

1. MANIFOLD STRUCTURE

1.1 *Topological vector space, differentiation, tensor spaces; exact sequences*

Given a real vector space V of dimension n we immediately have the dual vector space $V^* = L(V;\mathbb{R})$ of all real valued linear maps on V. The natural choice of topology for V is (cf. III §2.5)

$$T_V = \sup \{ f^{\leftarrow} T_{\mathbb{R}} \mid f \in V^* \}$$

where $T_{\mathbb{R}}$ is the standard metric topology for $\mathbb{R}$. It turns out that this topology for V has other equivalent formulations (cf. [28]). For example, we could take any basis $\{b_i \mid i = 1,2,\ldots,n\}$ for V and obtain a family of maps

$\{b^i \mid i = 1,2,\ldots,n\}$

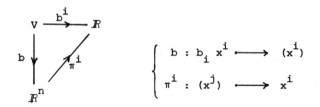

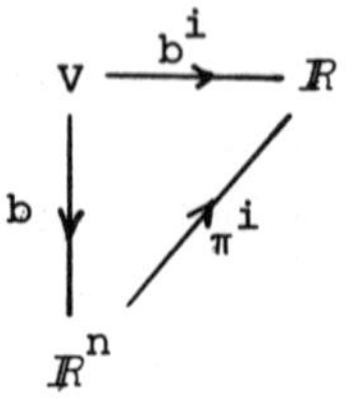

$$\begin{cases} b : b_i\, x^i \longmapsto (x^i) \\ \pi^i : (x^j) \longmapsto x^i \end{cases}$$

making the diagram commute. Here b is an isomorphism in Vec and the π^i are the projections from the product in Top. Evidently the b^i form a basis dual to the b_i. Since finite linear combinations of real valued maps are continuous we could topologise V by using

$$T'_V = \sup\{b^{i\leftarrow}T_{\mathbb{R}} \mid i = 1,2,\ldots,n\} .$$

However, $T'_V = T_V$ (cf. [28] for details and pictures of open sets). We call T_V the <u>usual topology</u> for V. Similarly, we shall always understand that vector spaces of linear maps, like $L(V;W)$, have the usual topology. This all fits together with the familiar situation of the linear maps from $\mathbb{R}^n$ to $\mathbb{R}^m$ being represented by matrices in $\mathbb{R}^{m\times n} = \mathbb{R}^{mn}$.

The <u>tangent space</u> at $x\in V$ is the vector space

$$T_xV = \{x\}\times V$$

which is isomorphic to V (in Vec and Top). Given any map (not necessarily linear)

$$f : V \longrightarrow W$$

between vector spaces V and W then the <u>derivative</u> of f at $x\in V$ (if it exists it is unique, cf. [28]) is a <u>linear map</u>

$$D_xf : T_xV \longrightarrow T_{f(x)}W$$

such that:

for any neighbourhood of $\underline{O} \in L(T_xV;\ T_{f(x)}W)$

there is a neighbourhood N' of $\underline{O} \in T_xV$

for which

$$(\forall(x,t)\in N')(\exists A\in N) : (f(x),f(x+t)) = D_xf(x,t) + A(x,t) .$$

Now, if $\dim V = n$, $\dim W = m$ then, once we choose bases for $T_x V \equiv V$ and $T_{f(x)} W \equiv W$, the linear map $D_x f$ is represented by a unique matrix $[\partial_i f^j]_x$ making the following diagram commute

$$T_x V \xrightarrow{\ D_x f\ } T_{f(x)} W$$

$$b \downarrow \qquad\qquad \downarrow c$$

$$\mathbb{R}^n \xrightarrow[\ [\partial_i f^j]_x\ ]{} \mathbb{R}^m \quad : \quad (t^i) \longmapsto (t^i \partial_i f^j) \ .$$

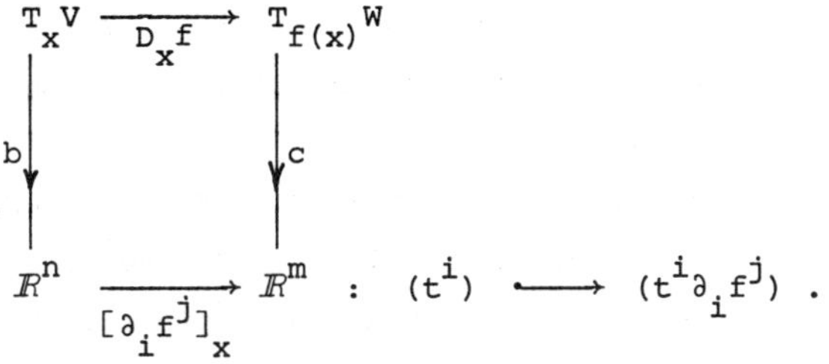

Of course, the entries in $[\partial_i f^j]_x$ are the partial derivatives the constituent maps of f via the chosen bases :

$$V \xrightarrow{\ f\ } W$$

$$\downarrow \qquad\qquad \downarrow$$

$$\mathbb{R}^n \longrightarrow \mathbb{R}^m \quad : \quad (x^i) \longmapsto \left(f^j(x^i) \right)$$

$$f^j \searrow \qquad \downarrow \pi^j$$

$$\mathbb{R}$$

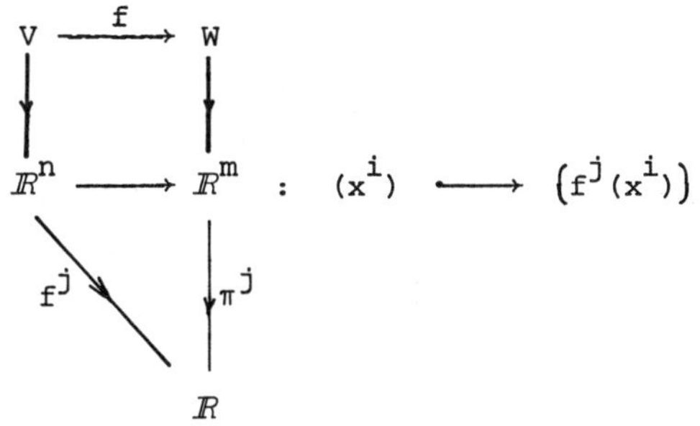

A map $f : V \longrightarrow W$ is <u>differentiable at x</u> if $D_x f$ exists, and f is called <u>differentiable</u> if $D_x f$ exists for all $x \in V$.

Since $T_x V$ and $T_{f(x)} W$ have their usual topologies we know that the linear map $D_x f$ is continuous. For handling higher derivatives it is useful to introduce (for differentiable f) the map

$$\hat{D}f \ : \ V \longrightarrow L(V;W)$$

$$: \ x \longmapsto \hat{D}_x f$$

where

$$\begin{array}{ccccccc} \hat{D}_x f : V & \longrightarrow & W & : & t & \longmapsto & t' \\ \downarrow & & \downarrow & & \downarrow & & \downarrow \\ T_x V & \xrightarrow{D_x f} & T_{f(x)} W & : & (x,t) & \longmapsto & (f(x),t') . \end{array}$$

Then we say : f is C^1 if $\hat{D}f$ is continuous; f is C^2 if $\hat{D}f$ is C^1, etc. If f is C^k for all finite k then we say that f is C^∞ or <u>smooth</u> .

For a detailed discussion of the above and for the more general definition of differentiation on affine spaces see [28] Ch. 7, and Porteous [78] Ch. 18. In [28] an appendix provides a modern proof of the inverse function theorem through the proof of a theorem on the existence and uniqueness of smooth flows (on a manifold). The classical form for the inverse function theorem for $f : \mathbb{R}^n \longrightarrow \mathbb{R}^n$ is:

1.1.1 <u>Theorem</u>

If f is C^k for any k then $D_x f$ is an isomorphism if and only if there are neighbourhoods N of x, N' of f(x) with f(N) = N' and there is a local C^k inverse $\overleftarrow{f} : N' \longrightarrow N$. □

<u>Corollary</u>

If $f : \mathbb{R}^n \longrightarrow \mathbb{R}^m$ $(n \leqslant m)$ is C^1 and $D_x f$ is injective then there is a neighbourhood N of x such that $f|_N$ is injective. □

In these assertions the domain of f can also be any non-empty set in $\mathbb{R}^n$; both carry over to maps between manifolds.

For physics it turns out that real valued linear functions of several vector variables are important, the <u>multilinear functionals</u> or <u>tensors</u>. A detailed motivation for the following definition will be found in [28] Ch. V. A <u>tensor product</u> of vector spaces $X_1, \ldots, X_n$ is a vector space X together with a multilinear map

$$\otimes : X_1 \times X_2 \times \ldots \times X_n \longrightarrow X$$

such that if also there is a multilinear map

$$f : X_1 \times X_2 \times \ldots \times X_n \longrightarrow Y$$

then there is a unique linear map

$$\hat{f} : X \longrightarrow Y \text{ with } \hat{f} = f \circ \otimes .$$

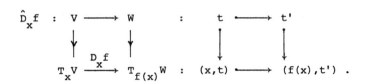

Diagrammatically (cf. II §2.2) we require commutativity of

$$\begin{array}{ccc} X_1 \times X_2 \times \ldots \times X_n & \xrightarrow{\quad \otimes \quad} & X \\ & {\scriptstyle f}\searrow \qquad \swarrow{\scriptstyle \hat{f}} & \\ & Y & \end{array}$$

It follows that the tensor product always exists and is unique up to isomorphism (cf. [28] p. 146). The actual construction of

$$X = X_1 \otimes X_2 \otimes \ldots \otimes X_n$$

is quite simple. First we define (cf. II §1.4 for the introduction of dual vector spaces):

$$X_1^* \otimes X_2^* \otimes \ldots \otimes X_n^* = L(X_1, X_2, \ldots, X_n; \mathbb{R}) ;$$

then we exploit the finite-dimensional natural isomorphism of taking double duals (cf. II §1.7) to put $X_i^{**} \equiv X_i$ and hence choose

$$X_1 \otimes X_2 \otimes \ldots \otimes X_n = L(X_1^*, X_2^*, \ldots, X_n^*; \mathbb{R}) .$$

Example

Given any vector spaces X_1, X_2 and their duals X_1^*, X_2^* , typical elements are

$$x \in X_1,\ y \in X_2,\ f \in X_1^*,\ g \in X_2^*$$

and then

$$x \otimes y : X_1^* \times X_2^* \longrightarrow \mathbb{R} : (f,g) \longmapsto f(x)\ g(y)$$

is an element of $X_1 \otimes X_2$ (and every element of $X_1 \otimes X_2$ is a linear combination of such individual products, though the map $\otimes$ so defined is not surjective, cf. [28]). Similarly,

$$f \otimes g : X_1 \times X_2 \longrightarrow \mathbb{R} : (x,y) \longmapsto f(x)\ g(y)$$

is an element of $X_1^* \otimes X_2^*$.

Formally, the tensor product is a bifunctor or functor of two variables (cf. II §2.7):

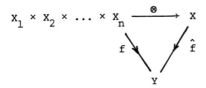

$$\mathrm{Vec} \times \mathrm{Vec} \longrightarrow \mathrm{Vec} \left\{ \begin{array}{ll} : & (A,B) \longmapsto A\otimes B \\ : & \begin{array}{c} A \quad B \\ f\downarrow \quad \downarrow g \\ A' \quad B' \end{array} \longmapsto \begin{array}{c} A\otimes B \\ \downarrow f\otimes g \\ A'\otimes B' \end{array} \end{array} \right.$$

where

$$f\otimes g \left(\sum_i a_i \otimes b_i\right) = \sum_i \left(f(a_i)\otimes g(b_i)\right) .$$

Of course this construction can equally well be performed in the category of Abelian groups. In general, the adjoint functor theorem (cf. Ch. II §2.8) provides for the construction of tensor products. Further discussion will be found in MacLane [67] p.159-160 and p.222.

Certain diagrams in the category Vec have particular significance, these are the <u>exact sequences</u>. The notion extends to other categories, notably to subcategories of Ab (cf. Herrlich and Strecker [46] §39 and MacLane [67] p.196) and, as we shall see below, to vector bundles (cf. Lang [60] p.48).

A diagram in Vec of the form

$$\cdots V_{-1} \xrightarrow[f_{-1}]{} V_o \xrightarrow[f_o]{} V_1 \xrightarrow[f_1]{} V_2 \cdots$$

(whether finite or infinite) is called a <u>sequence</u> and it is called an <u>exact sequence</u> if

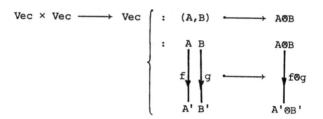

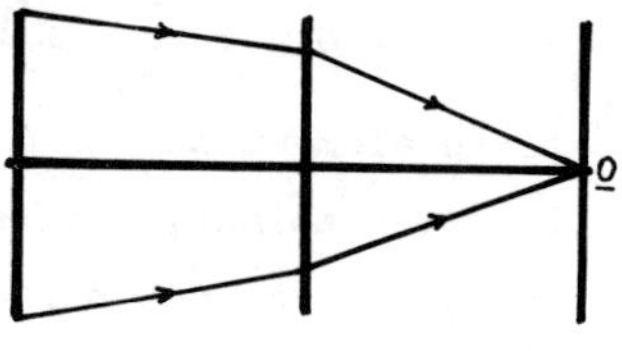

$$(\forall k) \ \mathrm{im}\, f_k = \ker f_{k+1} ;$$

this means

$$(\forall k) \quad f_k(V_k) = \overleftarrow{f}_{k+1}\{\underline{O}\} .$$

For any vector space V we usually omit names for the unique maps involving a trivial space:

$$\{\underline{O}\} \longrightarrow V \ : \ \underline{O} \longmapsto \underline{O}$$

$$V \longrightarrow \{\underline{O}\} : x \longmapsto \underline{O}$$

Examples

(Cf. Porteous [78] p.90-95 for basic theory.)

1. $\{\underline{0}\} \longrightarrow W \xrightarrow{s} X$ is exact $\Longleftrightarrow$ s injective .

2. $X \xrightarrow{t} Y \longrightarrow \{\underline{0}\}$ is exact $\Longleftrightarrow$ t surjective .

3. $\{\underline{0}\} \longrightarrow X \xrightarrow{t} Y \longrightarrow \{\underline{0}\}$ is exact

$\Longleftrightarrow$ t is an isomorphism

4. $\{\underline{0}\} \longrightarrow X \xrightarrow{i} X\times Y \xrightarrow{q} Y \longrightarrow \{\underline{0}\}$

and $\{\underline{0}\} \longrightarrow Y \xrightarrow{j} X\times Y \xrightarrow{p} X \longrightarrow \{\underline{0}\}$

(with i and j the obvious injections and q,p the obvious projections) are exact with $p\circ i = I_X$ and $q\circ j = I_Y$.

5. Given a vector subspace W of a vector space X ,

$$\{\underline{0}\} \longrightarrow W \hookrightarrow X \xrightarrow{\pi} X/W \longrightarrow \{\underline{0}\}$$

is exact. Recall that the factor space, or <u>quotient space</u> is defined by:

$$X/W = \{[x]_W | x\in X\} ,$$

$$[x]_W = \{x + w | w\in W\} = \pi(x) \quad (\forall x\in X) ,$$

$$\lambda[x]_W = [\lambda x]_W , \quad \forall \lambda\in\mathbb{R},$$

$$[x]_W + [y]_W = [x+y]_W .$$

An exact sequence of the form

$$\{\underline{0}\} \longrightarrow W \xrightarrow{s} X \xrightarrow{t} Y \longrightarrow \{\underline{0}\}$$

is called a <u>short exact sequence</u> and in view of Example 5 it is often helpful to think of W as a subspace of X , and of Y as X/W . Such a short exact sequence is said to be a <u>split exact sequence</u> if there is given a linear section of t (cf. II §1.10) :

$$t' : Y \longrightarrow X \quad \text{with} \quad t\circ t' = I_Y .$$

1.2 *Manifold definitions, universal cover*

A C^k manifold of dimension n is a Hausdorff topological space M with a collection of open sets

$$\{U_\alpha \mid \alpha \in A\}$$

in M and corresponding maps

$$\phi_\alpha : U_\alpha \longrightarrow \mathbb{R}^n$$

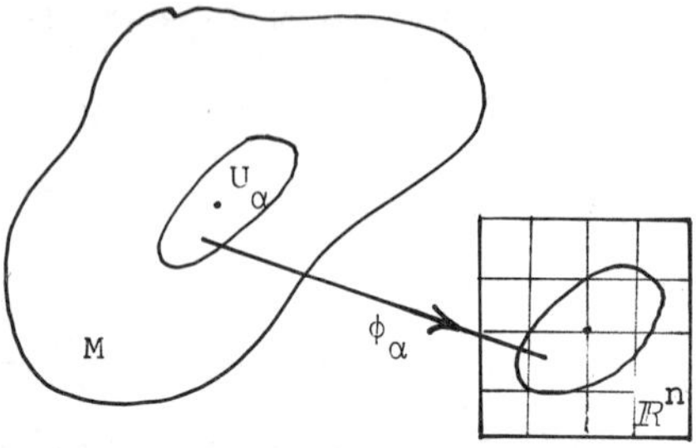

such that:

(i) $\bigcup_{\alpha \in A} U_\alpha = M$.

(ii) Each ϕ_α defines a homeomorphism $U_\alpha \longrightarrow \phi_\alpha(U_\alpha)$.

(iii) If $U_\alpha \cap U_\beta \neq \emptyset$ then the composites

$$\phi_\alpha \circ \overleftarrow{\phi_\beta} : \phi_\beta(U_\alpha) \longrightarrow \phi_\alpha(U_\beta)$$

$$\phi_\beta \circ \overleftarrow{\phi_\alpha} : \phi_\alpha(U_\beta) \longrightarrow \phi_\beta(U_\alpha)$$

are C^k maps (between subsets of $\mathbb{R}^n$) .

We call each pair (U_α, ϕ_α) a chart and the set $\{(U_\alpha, \phi_\alpha) \mid \alpha \in A\}$ an atlas for M . The properties of M are not significantly altered if we add to such an atlas more charts satisfying the requirements (i) - (iii) .

Unless we specify to the contrary we shall mean by 'manifold' a smooth (i.e. C^∞) manifold.

1.2.1 Theorem

Given a connected manifold M there is a unique universal covering manifold $\tilde{M}$; that is, a unique universal covering space (cf. III §1.7) with manifold structure.

Proof

Steenrod [96] pp. 67-71. □

Example

Real projective n-space $\mathbb{R}P^n$ is the quotient of the sphere

$$S^n = \{x \in \mathbb{R}^{n+1} \mid \|x\| = 1\}$$

by the group

$$G = \{I,J\} \quad \text{where} \quad \begin{cases} I : x \longmapsto x \\ J : x \longmapsto -x \,. \end{cases}$$

It turns out that $S^n = \widetilde{\mathbb{R}P}^n$, the sphere is the universal covering manifold of projective space.

1.3 Tangent space and derivatives : manifold category

A map $f : M \longrightarrow N$ between manifolds is called <u>differentiable</u> (or C^r) at $x \in M$ if for some charts (U,ϕ) on M, (V,ψ) on N with $x \in U$, $f(x) \in V$ the map

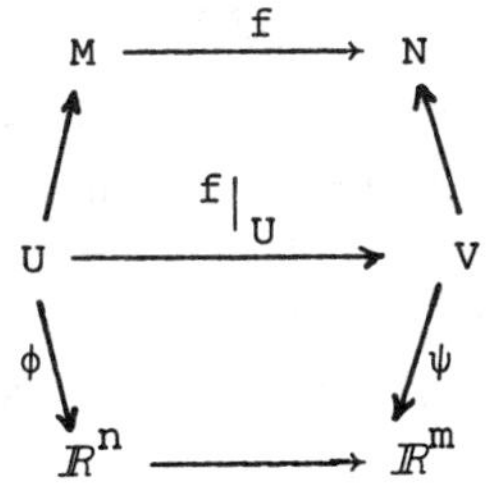

$$\psi \circ f|_U \circ \overleftarrow{\phi} : \phi(U) \longrightarrow \psi(V)$$

is differentiable (or C^r) at $\phi(x)$.
If f is differentiable (or C^r) at all $x \in M$ then we just call f <u>differentiable</u> (or C^r).

It turns out that this property is independent of the choice of charts at x and $f(x)$. Smooth manifolds and smooth maps between them form a category Man; isomorphisms in this category are called <u>diffeomorphisms</u>.

The <u>tangent space</u> to M at x is denoted T_xM and constructed as an isomorph in Vec of the tangent space $T_{\phi(x)}\mathbb{R}^n$; two equivalent constructions are discussed in detail in [28]. So if f as above is differentiable at x then we obtain a linear map, the <u>derivative</u> of f at x:

$$D_xf : T_xM \longrightarrow T_{f(x)}N \,.$$

Further, by means of the local co-ordinates induced by the chart maps

$$\phi : U \longrightarrow \mathbb{R}^n : x \longmapsto (x^1, x^2, \ldots, x^n)$$

$$\psi : V \longrightarrow \mathbb{R}^m : y \longmapsto (y^1, y^2, \ldots, y^m)$$

the map D_xf is uniquely determined by the Jacobian matrix $[\partial_i f^j]_x$, which is the derivative at $\phi(x)$ of

$$\psi \circ f|_U \circ \phi^{\leftarrow} : \mathbb{R}^n \longrightarrow \mathbb{R}^m : (x^i) \longmapsto (f^j(x^1, x^2, \ldots, x^n))$$

The <u>rank</u> of f at x is defined to be the rank of the linear map $D_x f$; it coincides with the matrix rank of $[\partial_i f^j]_x$.

If the rank of f at $x \in M$ is maximal then we say that x is a <u>regular point</u> of f , otherwise x is a <u>singular point</u> of f . The important result is due to Sard:

1.3.1 <u>Theorem</u>

If $f : M \longrightarrow N$ is smooth and S is the set of singular points of f then S has measure zero.

<u>Proof</u>

Schwartz [87] p.9. Note that S is said to have measure zero if for all charts (U,ϕ)

$\phi(U \cap S)$ has Lebesgue measure zero in $\mathbb{R}^n$. □

(Also in Schwartz [87] p.7, by the way, will be found proof that every connected one-dimensional manifold is diffeomorphic to either an open interval or a circle. These two results yield the Brouwer Fixed-Point Theorem.)

A <u>curve</u> in a manifold M is a map (C^1 unless otherwise stated)

$$c : I \longrightarrow M : t \longmapsto c(t)$$

from some interval $I \subseteq \mathbb{R}$. The <u>tangent vector to</u> c at $t \in I$ is denoted $\dot{c}(t)$ and defined by $\dot{c}(t) = D_t c\ (1)$, where 1 is the standard unit vector in $T_t\mathbb{R}$.

1.4 <u>Submanifold, product and quotient manifolds</u>

A subset S of a manifold M is a <u>submanifold</u> of M if the inclusion map $i : S \hookrightarrow M$ is an <u>immersion</u>, that is if the rank of i equals the dimension of S . This of course depends on providing first a manifold structure for S , the construction can be found for example in Lang [60] p.24.

The <u>product manifold</u> of two manifolds M,N is the product topological space $M \times N$ (cf. III §2.6) provided with the obvious differentiable structure from the products of chart maps. Evidently this process easily extends to any finite number of factors; we shall not need infinite product manifolds. The dimension of a product manifold is the sum of the dimensions of its constituent factors.

A differentiable map $f : M \longrightarrow N$ between manifolds is called a <u>submersion</u> if its rank equals the dimension of N at each point of its domain.
Then $(\forall x \in M)$ $D_x f$ is surjective.

Given an equivalence relation ρ on a manifold M we can take the set of equivalence classes

$$X = M/\rho = \{[x]_\rho \mid x \in M\}$$

and so obtain a surjection

$$g : M \longrightarrow M/\rho : x \longmapsto [x]_\rho \ .$$

If M/ρ is provided with a differentaible structure that makes g a submersion then we say that M/ρ is a <u>quotient manifold</u> of M. Often an equivalence relation arises on M from the action of a group, as we shall see below. Brickell and Clark [9] Ch. 6 discusses quotient manifolds in some detail and provides examples.

A number of manifolds arise as subsets of $\mathbb{R}^n$ for some n. The following theorem gives a large family of these.

1.4.1 <u>Theorem</u>

Let $f : \mathbb{R}^n \longrightarrow \mathbb{R}$ be smooth and let $M = \overleftarrow{f}\{0\}$. If at each point $x \in M$ f has rank one then M admits a C^∞ manifold structure of dimension n-1.

<u>Proof</u>

The construction of the atlas depends on the implicit function theorem and full details can be found in Brickell and Clark [9] p.25. □

The manifold so constructed is called a <u>differentiable variety</u> and the spheres are examples. As another example we have the set of all non-singular n×n matrices with real elements; this is evidently the differentiable variety (in $\mathbb{R}^{n^2}$) given by the determinant function. It is important because it is also a group with smooth operations.

1.5 *Lie group : action : transitive, free, effective*

A Lie group is a group G that is also a smooth manifold such that the group operation

$$G \times G \longrightarrow G : (a,b) \longmapsto ab^{-1}$$

is smooth. This means that the following maps are smooth for all $a \in G$

$$L_a : G \longrightarrow G : b \longmapsto ab \quad \text{(left translation)}$$

$$R_a : G \longrightarrow G : b \longmapsto ba \quad \text{(right translation)}$$

$$^{-1} : G \longrightarrow G : b \longmapsto b^{-1} \quad \text{(inversion)} .$$

We shall use e to denote the identity element in G.

Example 1

The general linear group $G\ell(n;\mathbb{R})$ of all $n \times n$ real non-singular matrices is an open submanifold of $\mathbb{R}^{n^2}$. We shall write $\mathbb{R}^* = G\ell(1;\mathbb{R})$.

Example 2

The group of real matrices:

$$\left\{ \begin{bmatrix} \beta & v\beta \\ v\beta & \beta \end{bmatrix} \text{ with } 1>v^2 \text{ and } \beta = (1-v^2)^{-\frac{1}{2}} \right\}$$

is an important Lie group in relativity, as a one-parameter subgroup of the Lorentz group (cf. Porteous [78] p.161). It has an equivalent representation as the matrix group: (cf. Gel'fand, Minlos and Shapiro [35] p.184 et seq.):

$$\left\{ \begin{bmatrix} \cosh\chi & \sinh\chi \\ \sinh\chi & \cosh\chi \end{bmatrix} \text{with } \chi \in \mathbb{R} \right\} .$$

Let G be a Lie group and P a manifold. Then G acts on P to (or on) the right if there is a map

$$P\times G \longrightarrow P : (u,g) \longmapsto R_g(u)$$

satisfying

(i) $(\forall g\in G)\ g : P \longrightarrow P : u \longmapsto R_g(u)$ is a diffeomorphism

(ii) $(\forall g,\ h\in G,\ \forall u\in P)\ R_{gh}(u) = R_h \circ R_g(u)$.

Example 1

Take $P = G$ and use right translation in G :

$$G\times G \longrightarrow G : (h,g) \longmapsto R_g(h) = hg .$$

An action of a Lie group G on the right of a manifold P is called

(i) transitive if $(\forall u,\ v\in P)\ \exists g\in G : R_g(u) = v$;

(ii) free if the only element of G with a fixed point is e ;

(iii) effective if

$$(R_g(u) = u\ \forall u\in P) \Longrightarrow g = e .$$

Example 2

Take $G = \mathbb{R}^*$, the multiplicative group on $\mathbb{R}\backslash\{0\}$ and $P = \mathbb{R}^2$ then

$$\mathbb{R}^2\times\mathbb{R}^* \longrightarrow \mathbb{R}^2 : \big((x,y),g\big) \longmapsto (gx,gy)$$

is not transitive. However, right translation in any G (cf. Example 1) is always transitive because

$$x,y\in G \Longrightarrow (\exists g = x^{-1}y) : R_g(x) = y .$$

Example 3

Let $P = S^1\times\mathbb{R}^*$. Then the action induced on P by right translation in $\mathbb{R}^*$ is free. However, the action of $\mathbb{R}^*$ on $\mathbb{R}^2$ in Example 2 is not free because every element of $\mathbb{R}^*$ has a fixed point at the origin.

Given a Lie group G acting on the right of a manifold P, the orbit of G through $u\in P$ is the set

$$[u]_G = \{R_g(u)\,|\,g\in G\}$$

and we denote by P/G the set of all such orbits for $u\in P$. This factor space is topologized by requiring that the projection

$$\Pi_P : P \longrightarrow P/G : u \longmapsto [u]_G$$

is continuous.

We shall denote the connected component of the identity in a Lie group G by G^+. The topology on G is second countable if and only if the quotient group G/G^+ is countable (or finite). By a Lie subgroup H of a Lie group G we shall mean a subgroup that is also a submanifold and a Lie group with respect to this structure; we shall also require H to be second countable to ensure uniqueness (cf. [57] p.40).

Example 4

When $G = G\ell(1;\mathbb{R}) = \mathbb{R}^*$ we find

$$G^+ = G\ell(1;\mathbb{R})^+ = \mathbb{R}^+ = \{g\in\mathbb{R} \mid g>0\} .$$

An illustrated study of the geometry of the Lie group $S\ell(2;\mathbb{R})$, consisting of unimodular operators on $\mathbb{R}^2$, is given in Dodson and Poston [28]. Lie groups are special cases of topological groups, for a study of the category of the latter see Higgins [48].

Next we collect some important general results.

1.5.1 Theorem

Let G be a locally compact topological group, H a Lie group and $h : G \longrightarrow H$ a continuous homomorphism with h injective on some neighbourhood of the identity in G.

Then G is a lie group.

Proof

Hochschild [50] Ch. 8. □

Corollary

Every closed subgroup of a Lie group is locally compact and therefore a Lie group. □

1.5.2 Theorem

Let G be a simply connected, connected Lie group and H a normal, connected Lie subgroup. Then H is closed in G. Also, the canonical map

$$\pi_H : G \longrightarrow G/H$$

is smooth, and there exists a smooth map

$$\rho_H : G/H \longrightarrow G$$

such that $\pi_H \circ \rho_H = I_{G/H}$ and

$$H\times(G/H) \longrightarrow G : (x,u) \longmapsto x\rho_H(u)$$

is an isomorphism in Man.

Proof

Hochschild [50] p.135. Note that H and G/H are necessarily simply connected. □

1.5.3 Theorem

Let H be a connected Lie subgroup of a connected Lie group G . If the closure in G of every one-dimensional connected Lie subgroup of H lies in H , then H is closed in G .

Proof

Hochschild [50] p.192. □

1.5.4 Theorem

Every continuous homomorphism of a connected Lie group into a connected Lie group is smooth.

Proof

Hochschild [50] p.84. □

1.5.5 Theorem

Any topological group which has a C^1-Lie group structure admits a unique C^∞-Lie group structure compatible with the given C^1 structure.

Proof

Pontryagin [77]. □

Remark

Data on the commoner Lie groups (general linear, special linear, orthogonal, unitary, special orthogonal, special unitary, symplectic and spin) can be found in Porteous [78] p.421 et seq. For more of the general theory see Chevalley [12] and Kobayashi [56], for detailed applications see DeWitt and DeWitt [23].

2. STRUCTURES ON MANIFOLDS

2.1 *Vector bundle, exact sequence*

We have seen in §1.3 that local tangent vector spaces arise naturally on (smooth) manifolds. These vector spaces can be fitted together to form a new manifold, the tangent bundle. More generally, we can use other vector spaces to make arbitrary vector bundles over a given manifold. Later we obtain tangent bundles via a functor from the category of manifolds to the category of vector bundles. We outline the construction of the category of vector bundles, following mainly the development in Lang [60] Ch. 2. Throughout, our manifolds and maps will be assumed to be smooth.

A vector bundle with fibre V over a manifold M is a triple $(M.E,\Pi)$ where E is a manifold, V is a vector space and

$$\Pi : E \longrightarrow M$$

is a surjection satisfying:

(i) $(\forall x \in M)\, \overleftarrow{\Pi}(x)$ is isomorphic to V in Vec.

(ii) $(\forall x \in M)$ ($\exists$ open U containing x) and a diffeomorphism

$$\tau : \overleftarrow{\Pi}(U) \longrightarrow U \times V$$

commuting with the projections

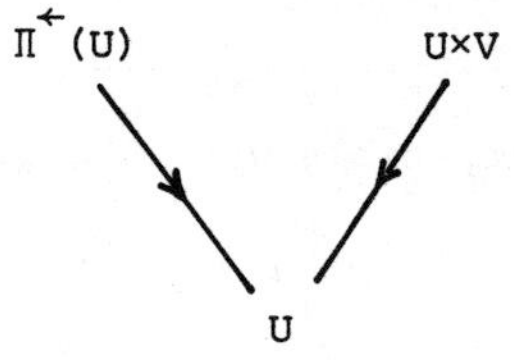

and the induced maps on fibres

$$\tau_x : \overleftarrow{\Pi}(x) \longrightarrow V$$

are isomorphisms in Vec. (Since we are in finite dimensions the

τ_x are also isomorphisms in Top because the usual topology always makes linear maps continuous.)

We call M the <u>base space</u>, E the <u>total space</u> and Π the bundle map of (M,E,Π). The class of vector bundles becomes a category, V bun, with morphisms defined as follows.

A <u>vector bundle morphism</u> between two vector bundles (M,E,Π), (M',E',Π') is a pair of maps

$$h_o : M \longrightarrow M' \ , \quad h : E \longrightarrow E'$$

satisfying:

$$\begin{array}{ccc} E & \xrightarrow{\ h\ } & E' \\ \Pi \big\downarrow & \text{commutes} & \big\downarrow \Pi' \\ M & \xrightarrow[\ h_o\ ]{} & M' \end{array} \qquad \left\{ \begin{array}{l} \text{and each restriction} \\ \\ h_x : \Pi^{\leftarrow}(x) \longrightarrow \Pi'^{\leftarrow}\big(h(x)\big) \\ \\ \text{is a continuous linear map.} \end{array} \right.$$

In applications we often use several vector bundles over a given manifold; then h_o is the identity.

In Lang [60] will be found fuller descriptions, extended to the case of infinite dimensional manifolds with vector bundles modelled on Banach spaces, and details of operations on vector bundles. We note that unlike Man, the category V bun does admit direct sums (the Whitney sum), quotients and pullbacks (cf. Hirsch [49] for specific constructions). Functional operations available in Vec such as tensor products and exact sequences can also be carried over to V bun (cf. Lang [60] Ch. 3). Vector bundles over a given manifold M form a subcategory V bun(M) (cf. II §1.5). There is a zero object in V bun(M) (cf. II §1.9), the <u>zero-dimensional bundle over M</u>, denoted $\{\underline{0}\}$. We shall look next at exact sequences, but first we point out that these, like the constructions above, will be for smooth vector bundles over smooth manifolds. The whole thing could be presented for C^r-bundles over C^k-manifolds but then care is necessary when differential processes arise and certain operations cease to be closed. In any <u>particular</u> application it is not difficult to make the appropriate substitution of known classes of differentiability; it is tedious to do so generally.

The idea of exact sequences (cf. §1.1) can be transferred to vector bundles (cf. Lang [60] p. 48 and Hirsch [49] p. 93). Given a manifold M and two vector bundles (M,E,Π), (M,E',Π') over M with fibres V and V' respectively then a diagram (in V bun)

$$\{\underline{O}\} \longrightarrow E' \xrightarrow{f} E$$

is called an <u>exact sequence</u> if f is injective and M has an open cover K such that for each $U \in K$ there is a commutative diagram in Man:

$$\begin{array}{ccc} \Pi'^{\leftarrow}U & \xrightarrow{\quad f \quad} & \Pi^{\leftarrow}U \\ \big\downarrow{\scriptstyle \tau'} & & \big\downarrow{\scriptstyle \tau} \\ U\times V' & \xrightarrow[(I_U,\, I_{V'},\, \underline{O})]{} & U\times V'\times W \;=\; U\times V \end{array}$$

where the τ' and τ are diffeomorphisms.

Now we can define $S \subset E$ to determine a <u>subbundle</u> of (M,E,Π) if there is an exact

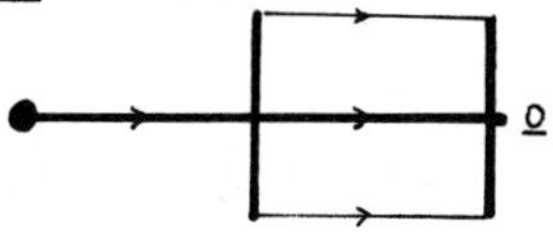

$$\{\underline{O}\} \longrightarrow E' \xrightarrow{f} E$$

with $f(E') = S$ (cf. Examples in §1.1).

Similarly, a diagram (in V bun)

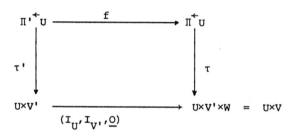

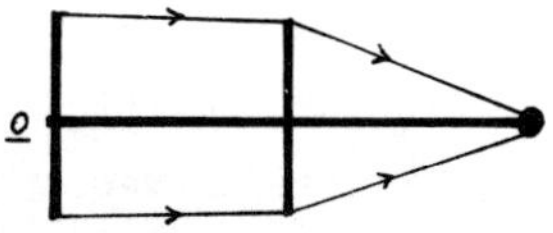

$$E \xrightarrow{g} E'' \longrightarrow \{\underline{O}\}$$

is an <u>exact sequence</u> if g is surjective and M has an open cover K such that for each $U \in K$ there is a commutative diagram in Man:

$$\begin{array}{ccc} \Pi^{\leftarrow}U & \xrightarrow{\quad g \quad} & \Pi''^{\leftarrow}U \\ \big\downarrow{\scriptstyle \tau} & & \big\downarrow{\scriptstyle \tau''} \\ U\times V = U\times V'\times W & \xrightarrow[(I_U,\, \Pi_W)]{} & U\times W \end{array}$$

where the τ and τ'' are diffeomorphisms.

We can tie these definitions in with those in Vec (cf. §1.1) by defining for any vector bundle morphism (cf. §2.1)

$$\begin{array}{ccc} E & \xrightarrow{h} & E' \\ \Pi \downarrow & & \downarrow \Pi' \\ M & \xrightarrow[h_o]{} & M' \end{array} \qquad \left\{ \begin{array}{l} (\forall x \in M) \text{ the restriction} \\ h_x : \Pi^{\leftarrow}(x) \longrightarrow \Pi'^{\leftarrow}(h(x)) \\ \text{is linear.} \end{array} \right.$$

$$\ker(h) = \bigcup_{x \in M} \ker(h_x) \qquad \text{(the \underline{kernel} of h)}$$

$$\mathrm{im}(h) = \bigcup_{x \in M} \mathrm{im}(h_x) \qquad \text{(the \underline{image} of h)} .$$

The kernel and image of a vector bundle morphism naturally inherit vector bundle structures because kernels and images of linear maps are necessarily vector subspaces (cf. Lang [60] §3 for further details). Now we can say that a diagram in V bun of the form

$$\cdots E_{-1} \xrightarrow[f_{-1}]{} E_o \xrightarrow[f_o]{} E_1 \xrightarrow[f_1]{} E_2 \cdots$$

is called an exact sequence if

$$(\forall k)\ (\forall x \in E_k)\ \mathrm{im}\, f_{kx} = \ker f_{k+1\, f_k(x)} .$$

Then a short exact sequence is one of the form

$$\{\underline{o}\} \longrightarrow E' \xrightarrow{f} E \xrightarrow{g} E'' \longrightarrow \{\underline{o}\}$$

and then f is a monomorphism, g is an epimorphism (cf. II §1.8) and $\mathrm{im}\, f = \ker g$. Such a short exact sequence is called split if there is a monomorphism $h : E'' \longrightarrow E$ such that $g \circ h = I_{E''}$ (cf. II §1.10). As before (cf. §1.1) we can view E'' in the short exact sequence as a quotient object of the monomorphism f, this quotient bundle is as usual determined up to isomorphism. In the particular case that $E' \subset E$ is a sub bundle then the quotient bundle is denoted E/E' and its fibres are the quotient vector spaces $\Pi^{\leftarrow}(x)/\Pi'^{\leftarrow}(x)$.

We pointed out earlier that V bun admits pullbacks (cf. II

§2.2). Given a vector bundle (M,E,Π) and a manifold map $f : M_o \longrightarrow M$ the induced bundle or pullback over M_o is

$$(M_o, f^*E, \Pi_o) \text{ with } f^*E = \{(x,y) \in M_o \times E \mid f(x) = \Pi(y)\}$$

and

$$\Pi_o : f^*E \longrightarrow M_o : (x,y) \longmapsto x .$$

The atlas for f^*E is the maximal one containing all charts

$$\psi : \overleftarrow{f} U \longrightarrow \mathbb{R}^n$$

for which there is a chart on E,

$$\phi : U \longrightarrow \mathbb{R}^n$$

satisfying

$$z \in \overleftarrow{f} U,\ f(z) = x \in U \Longrightarrow \psi|_{\overleftarrow{\Pi_o}(z)} = \phi|_{\overleftarrow{\Pi}(x)} .$$

2.1.1 Theorem (Covering homotopy theorem)

Let B be paracompact and suppose that E is a vector bundle over $B\times[0,1]$. Then E is isomorphic to the vector bundle $E_B\times[0,1]$ where:

$$E_B = E|_{B\times\{0\}} .$$

Proof

Hirsch [49] p. 90. □

2.1.2 Theorem

Let B be paracompact, $f,g : B \longrightarrow M$ homotopic maps and let (M,E,Π) be a vector bundle. Then f^*E is isomorphic to g^*E in V bun . In particular if g is constant then f^*E is trivial.

Corollary

Every vector bundle over a contractible paracompact space is trivial.

Proof

Hirsch [49] p. 97. (See III §1.7 for the definition of homotopy and contractibility.) □

We conclude this section with some applications of the above (cf. Hirsch [49] p. 94 et seq.). The Whitney sum of bundles E', E over M is the bundle $E' \oplus E$ with fibres, $\overleftarrow{\Pi'}(x) \oplus \overleftarrow{\Pi}(x)$ for $x \in M$. Direct sums of charts for E', E give charts for $E' \oplus E$ and the exact sequences of fibres (in Vec)

$$\{\underline{0}\} \longrightarrow \overleftarrow{\Pi'}(x) \longrightarrow \overleftarrow{\Pi'}(x) \oplus \overleftarrow{\Pi}(x) \longrightarrow \overleftarrow{\Pi}(x) \longrightarrow \{\underline{0}\}$$

can be collected to give a (split) exact sequence (in V bun)

$$\{\underline{0}\} \longrightarrow E' \longrightarrow E' \oplus E \longrightarrow E \longrightarrow \{\underline{0}\} .$$

2.1.3 Theorem

Every short exact sequence of vector bundles over a paracompact manifold M is split.

Proof

We seek a left inverse (cf. II §1.10) of a monomorphism $f : E' \longrightarrow E$. Locally this is available since $f(E')$ is a subbundle of E and there are open $U \subseteq M$ with charts

$$\overleftarrow{\Pi'}U \longrightarrow U \times \mathbb{R}^n \longrightarrow \mathbb{R}^m \times \mathbb{R}^n$$

$$\overleftarrow{\Pi}U \longrightarrow U \times \mathbb{R}^k \longrightarrow \mathbb{R}^m \times \mathbb{R}^k .$$

The inverse then appears as a linear map $\mathbb{R}^n \longrightarrow \mathbb{R}^k$. These local left inverses are then glued together by a partition of unity, available on M by its paracompactness (cf. III §1.6). □

2.1.4 Theorem

For every vector bundle E there exists a vector bundle E', over the same paracompact base M, such that $E \oplus E'$ is trivial.

Proof

The technique is again to use a partition of unity and to construct $E \oplus E' = M \times \mathbb{R}^m$, where m is the (finite) number of generators of the module of sections of the bundle E. Details will be found in Greub, Halperin and Vanstone [40] p. 76 et seq. □

2.1.5 Theorem

A vector bundle E over M, with fibre V and $\dim V = k$, is trivial if and only if there exist k sections

$$\sigma_i : M \longrightarrow E : x \longmapsto \sigma_i(x)$$

such that $(\forall x \in M)$

$$\{\sigma_i(x) \mid i = 1,2,\ldots,k\} \text{ is linearly independent.}$$

Proof

Such a family of sections gives an isomorphism with the trivial bundle $M \times \mathbb{R}^k$. □ (Cf. also Theorem 3 in §2.3 below.)

2.2 *Tangent bundle, differential, jet*

The tangent bundle functor T is a covariant functor from Man to V bun defined by

$$\begin{array}{ccc} M & & TN \\ \downarrow f & \overset{T}{\longmapsto} & \downarrow Df \\ M' & & TM' \end{array}$$

where (cf. II §2.2):

(i) $TM = (M, \coprod_{x \in M} T_x M, \Pi_T)$.

(ii) Π_T sends each $y \in T_x M$ to x ,

(iii) the commuting diagram for Df is

$$\left.\begin{array}{ccc} TM & \xrightarrow{Df} & TM' \\ \Pi_T \downarrow & & \downarrow \Pi'_T \\ M & \xrightarrow{f} & M' \end{array}\right\} \quad \text{and } Df|_{T_x M} = D_x f .$$

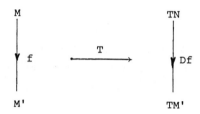

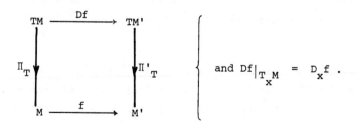

Thus, if M is an n-manifold then TM is a vector bundle with fibre $\mathbb{R}^n$. The bundle morphism Df is called the differential of f . Some authors denote Df by f_* . For full details of the synthesis of the differentiable structure on TM from the atlas for M see Dodson and Poston [28].

The results of the previous section on vector bundles have various particular applications to the tangent bundle. Firstly, from Lang [60] p. 53, given a manifold immersion (cf. §1.4)

$$f : M \longrightarrow M' \qquad \text{(in Man)}$$

we have a differential (in fact a monomorphism)

$$Df : TM \longrightarrow TM' \qquad \text{(in Man)}$$

but also a canonical vector bundle morphism

$$T^*f : TM \longrightarrow f^*TM', \quad \text{(in } \mathcal{V}\text{bun)}$$

by the universal property of pullbacks (cf. II §2.3).
Then there is an exact sequence

$$\{\underline{0}\} \longrightarrow TM \xrightarrow{T^*f} f^*TM' .$$

We can also work with submersions. In particular for a vector bundle (M,E,Π) we obtain the exact sequence (for $D\Pi$ is epic here)

$$\{\underline{0}\} \longrightarrow \Pi^*E \longrightarrow TE \longrightarrow \Pi^*TM \longrightarrow \{\underline{0}\} .$$

Next, from Hirsch [49] p. 94, given a vector bundle (M,E,Π) we can identify each vector space $\overleftarrow{\Pi}(x)$ with its own tangent space $T_x\overleftarrow{\Pi}(x)$.

Hence, E is a sub bundle of T_ME , the restriction of TE to $M \subset E$, well defined since M is isomorphic to the zero section of E . But then TM is also a sub bundle of T_ME via the differential of the inclusion $i : M \hookrightarrow E$. It follows that there is a short exact sequence

$$\{\underline{0}\} \longrightarrow E \longrightarrow T_ME \xrightarrow{D\Pi} TM \longrightarrow \{\underline{0}\}$$

which is split by

$$Di : TM \longrightarrow T_ME .$$

In consequence there is a natural isomorphism

$$T_ME \cong E \oplus TM$$

with the Whitney sum (cf. §2.1).

There are many ways to view tangent vectors on a manifold: equivalence classes of (C^1) curves going through a given point in a given direction; derivations on the algebra of (smooth) real functions on the manifold; equivalence classes of (C^1) real functions agreeing up to the first derivative; as geometrical directions in $\mathbb{R}^{2n}$, by virtue of the Whitney embedding theorem for manifolds with dimension n (and second countable topology, cf. III §1.5).

It is the aspect of 'differential agreement up to order one' that suggests the generalisation to jets. A tangent vector is a 1-jet or a jet of order 1; we proceed to construct n-jets. Denote by T^oM the (infinite-dimensional) vector space of smooth real functions on a manifold M. (The notation is chosen to be consistent with that for general tensor fields in the next section.) Now we follow Schwartz [87] p.2 and define:

> $g \in T^oM$ is n-horizontal at $x \in M$ if, in the co-ordinates of some (hence any) chart, all partial derivatives of g up to and including order n vanish at x.

Then an n-jet to M at x is a linear functional $\tau \in (T^oM)^*$ such that

$$g \in T^oM \text{ is n-horizontal} \Longrightarrow \tau(g) = 0 .$$

Now we find local co-ordinate expressions for n-jets.

Let

$$\phi : U \longrightarrow \mathbb{R}^n : \alpha \longmapsto (\alpha^1, \alpha^2, \ldots, \alpha^n)$$

be a chart about $x \in M$. Hence any $g \in T^oM$ has a local expression

$$g|_U \circ \overleftarrow{\phi} : \mathbb{R}^n \longrightarrow \mathbb{R} : (\alpha^i) \longmapsto \tilde{g}(\alpha^i) = g(\alpha)$$

and, in fact near $\underline{0} \in \mathbb{R}^n$,

$$\tilde{g}(\alpha^i) = C + \rho(\alpha^i) + H_n(\alpha^i)$$

where C is some constant, $H_n(\alpha^i)$ is the n-horizontal part and $\rho(\alpha^i)$ is the so-called principal part. For any n-jet τ at $x \in M$, by construction,

$$\tau(H_n(x^i)) = 0 = \tau(C)$$

and so, by the linearity of τ

$$\tau(\tilde{g}(x^i)) = \tau(\rho(x^i))$$

$$= \tau\left(\sum \partial_i \tilde{g}|_o x^i + \ldots \frac{1}{n!}\sum \partial_{i_1} \ldots \partial_{i_n} \tilde{g}|_o x^{i_1} \ldots x^{i_n}\right)$$

$$= \sum \partial_i \tilde{g}|_o \tau(x^i) + \ldots \frac{1}{n!}\sum (\partial_{i_1} \ldots \partial_{i_n} \tilde{g}|_o)\tau(x^{i_1} \ldots x^{i_n}).$$

Now we define the co-ordinates of τ by

$$\tau(x^i) = \tau^i, \ldots, \tau(x^{i_1} \ldots x^{i_n}) = \tau^{i_1 \ldots i_n}.$$

Hence a chart about $x \in M$ yields a chart about τ.
So there is an unambiguous representation

$$\tau(g) = \sum \partial_i \tilde{g}|_o \tau^i + \ldots + \frac{1}{n!}\sum \partial_{i_1 \ldots i_n} \tilde{g}|_o \tau^{i_1 \ldots i_n}.$$

In each term the summation is from $i=1$ to $i=\dim M$. If we use the summation convention and write the co-ordinates before the derivatives, omitting the function g we have

$$\tau = \tau^i \partial_i + \ldots + \frac{1}{n!}\tau^{i_1 \ldots i_n} \partial_{i_1 \ldots i_n}.$$

Then we easily recognise the case of the 1-jet as a tangent vector

$$\tau = \tau^i \partial_i \in T_x M,$$

that is a derivation on smooth real functions.

Evidently the n-jets at $x \in M$ admit finite linear combinations over $\mathbb{R}$ and so we have the n-jet vector space $J^n_x M$ to M at x. These spaces collect into a vector bundle (cf. Palais [74]), glued together via the above charts and those on M, the n-jet bundle $J^n M$. Sections of $J^n M$ yield n-jet fields and these with pointwise addition form an (infinite-dimensional) vector space in a similar way to the tensor fields. Moreover, there is a natural composition of jet fields arising from the natural composition of partial derivations in coordinates. So if τ is an n-jet field and σ is an m-jet field then $\tau\sigma$ is an (n+m)-jet field. For more details of the theory see Yano and Ishihara [107] where chapter X is particularly useful for 2-jet bundles.

(We note in passing that the manifold structure on $J^n M$ is of class C^k if M is of class C^{n+k}. Another generalisation, discussed in Hirsch [49], is the notion of n-jets from M to N where N is some manifold. These are equivalence classes of maps $M \longrightarrow N$ agreeing up to their n-th derivative.)

An integral curve of a (tangent) vector field w on M is a curve c in M such that (cf. §1.3)

$$(\forall t \in \text{dom } c) \; \dot{c}(t) = w \circ c(t) \; .$$

Such curves always exist for smooth vector fields and they are essentially unique. (cf. Dodson and Poston [28] Appendix for a recent geometric proof.) Intuitively such curves are found by 'joining up the arrows' given by the vector field as a tangent vector at each point. The process allows a generalization to 'joining up subspaces' in the following way (cf. Kobayashi and Nomizu [57] p. 10, Brickell and Clark [9] ch. 11).

A distribution S of dimension r on M is an assignment to each $x \in M$ of an r-dimensional subspace S_x of T_xM. We shall consider only those distributions that are differentiable, that is every $x \in M$ has a neighbourhood U on which there are r vector fields forming a basis for S_y, $y \in U$. So we can think of S as a 'field of tangent subspaces'. A vector field w on M is said to belong to S if

$$(\forall x \in M) \; w(x) \in S_x \; .$$

A distribution S is called involutive if : whenever w, v belong to S then so does $[w,v]$, and here $[\ ,\ ]$ is the Lie bracket or commutator :

$$[w,v]f = w(v(f)) - v(w(f)) \qquad (\forall f \in T^{0}M) \; .$$

A connected submanifold N of M is called an integral manifold of a distribution S if

$$(\forall x \in N) \; D_x i \; (T_x N) = S_{i(x)}$$

where i is the inclusion $N \hookrightarrow M$. Such an N is called maximal if no other integral manifold of S contains it. The generalization of the integral curve result is due to Frobenius (cf. Chevalley [12] for proof):

2.2.1 Theorem

Let S be an involutive distribution on a manifold M. Through every $x \in M$ there passes a unique maximal integral manifold $N(x)$ of S. Any integral manifold through x is an open submanifold of $N(x)$. □

We shall make use of distributions in two contexts; one arises from a connection (cf. §2.6 below) and the other arises from a Lie algebra (cf. §2.5 below). In the latter case the distribution is naturally involutive and indeed these play an important role in modern geometric quantization theory (cf. e.g. Simms and Woodhouse [89]). Another result is the following, also proved in Chevalley [12] (cf. [57] p. 11).

Proposition

Let S be an involutive distribution on a manifold M. Let N be a submanifold of M whose connected components are all integral manifolds of S. Given a manifold map $f : M' \longrightarrow M$ with $fM' \subseteq N$ then f is a manifold map from M' into N if N is second countable. □

2.3 *Fibre bundle : principle, frame, associated.*

A useful survey on fibre bundles is given in Eells [31].

We saw that vector bundles arise naturally over manifolds. More generally, we have the concept of a bundle (B,X,ρ) where

$$\rho : B \longrightarrow X \quad \text{(in Top)}$$

is a continuous surjection and B is viewed as a union of fibres $\{\rho^{\leftarrow}(x) \mid x \in X\}$ glued together via the topology of X. In our applications we usually have more structure for the fibres, usually of an algebraic character, these are the fibre bundles resulting from actions of groups. The simplest case is of vector bundles, where the fibre algebra is the additive group $\mathbb{R}^n$ for some n. Next, we wish to admit more general, not necessarily Abelian, group structure for fibres. However, we do not wish to lose the compatibility of the algebra with the topology so we need topological groups and in the case that our spaces are manifolds we shall demand compatibility of the algebra with the differential structure so we need Lie groups (cf. §1.5). Many of the bundles we shall use actually arise as associated bundles of a bundle of the following type.

A principal fibre bundle over a manifold M is a triple (P,G,M) where P is a manifold and G is a Lie group such that:

(i) G acts freely on P to the right,

(ii) $M = P/G$ and the canonical projection

$$\Pi_P : P \longrightarrow M : u \longmapsto [u]$$

is smooth,

(iii) P is locally trivial; that is, every $x \in M$ has a neighbourhood U such that $\overleftarrow{\Pi}_P U$ is diffeomorphic to $U \times G$.

We call G the structure group of the bundle; property (ii) ensures a transitive action on all fibres $\overleftarrow{\Pi}_P(x)$.

The vertical subspace of the tangent space T_uP for $u \in P$ is the kernel of $D\Pi_P$:

$$G_u = \{X \in T_uP \mid D\Pi_P X = 0 \in T_{\Pi_P(u)}M\} .$$

Example 1

The trivial product bundle, $P = M \times G$.

Example 2

The universal covering manifold $\tilde{M}$ (cf. §1.2) is a principal fibre bundle over M with structure group $\pi_1(M)$, the fundamental group of M (cf. Singer and Thorpe [90] and ch. III §1.7). Note that Chevalley [12] called this group the Poincaré group.

Example 3

The frame bundle over a manifold M :

$$P = LM = \{(x,(X_i)) \mid x \in M;\ (X_i) \text{ an ordered basis for } T_xM\}$$

$$G = G\ell(n;\mathbb{R}),$$

where n is the dimension of M .

This example is important in applications for we can derive from it all of the <u>tensor bundles</u>. We see that given any basis for T_xM at $x \in M$ (e.g. by choosing a chart about x) we obtain an isomorphism

$$T_xM \longrightarrow \mathbb{R}^n : \alpha \longmapsto (\alpha^i)$$

which allows $G\ell(n;\mathbb{R})$ to act by matrix multiplication on the coordinates (α^i) . Hence the orbit through $(x,(X_i)) \in LM$ is given by

$$[(x,(X_i))] = \{(x,(X_j g^j_i)) \mid (g^j_i) = g \in G\ell(n;\mathbb{R})\} .$$

But as g runs through $G\ell(n;\mathbb{R})$, so $R_g(x,(X_i))$ runs through all bases for T_xM . Hence, we may as well abbreviate $[(x,(X_i))]$ to x , so obtaining the projection map

$$\Pi_L : LM \longrightarrow M : (x,(X_i)) \longmapsto x .$$

We can see how LM and TM are related by the tangent bundle functor (cf. §2.2) as follows:

The canonical projections

$$\Pi_T : TM \longrightarrow M, \ \Pi_L : LM \longrightarrow M$$

yield an exact sequence (cf. §2.1)

$$\{\underline{0}\} \longrightarrow \text{kernel } (T\circ\Pi_L) \longrightarrow TLM \longrightarrow \Pi_L^*(TM) \longrightarrow \{\underline{0}\}$$

and it follows that kernel $(T\circ\Pi_L)$ and the pullback (cf. §2.1) $\Pi_L^*(TM)$ are trivial vector bundles.

The <u>canonical one-form</u> of a frame bundle LM is the map

$$\theta : TLM \longrightarrow \mathbb{R}^n$$

$$: (u,W) \longmapsto \Pi_u \circ D\Pi_L(u,W)$$

where for all $u = (x,(u_i)) \in LM$

$$\Pi_u : T_xM \longrightarrow \mathbb{R}^n$$

$$: \alpha^i u_i \longmapsto (\alpha^i) .$$

We note that, by construction, vertical vectors lie in the kernel of $D\Pi_L$ so $\theta(G_u) = \underline{0}$ for all u . In components we see that θ is a matrix action:

$$\theta : TLM \longrightarrow \mathbb{R}^n$$

$$: (x^i, b^i_j, X^i, B^i_j) \longmapsto [b^i_j]^{-1}\,[X^i] \;.$$

Here, $(x^i) \in \mathbb{R}^n$ are coordinates of $x \in M$ with respect to some chart (U,ϕ) and these determine basis fields $(\partial_i)_{i=1,\ldots,n}$ with respect to which a frame u is given by

$$(b^i_j \partial_i)_{j=1,\ldots,n} = (u_j)_{j=1,\ldots,n} \;.$$

The X^i are components with respect to $(\partial_i)_{i=1,\ldots,n}$ of a typical $X \in T_xM$. Evidently we have, using the summation convention,

$$X = X^i \partial_i = Y^j u_j = Y^j b^i_j \partial_i$$

for some $[Y^j] = [b^i_j]^{-1}[X^i] \in \mathbb{R}^n$. So θ gives the components of X with respect to the frame u .

Example 4

The trivial product bundle over S^1, $LS^1 = S^1 \times \mathbb{R}^*$ has two components, corresponding to positively oriented bases

$$L^+S^1 = S^1 \times \mathbb{R}^+$$

and to negatively oriented bases

$$L^-S^1 = S^1 \times \mathbb{R}^- \;.$$

Quite generally, LM has two components if M is orientable otherwise LM is connected. Precisely, a manifold M with atlas $\{(U_\alpha, \phi_\alpha) \mid \alpha \in A\}$ is orientable if there is a sub-atlas $\{(U_\alpha, \phi_\alpha) \mid \alpha \in A'\}$ such that whenever $U_\alpha \cap U_\beta \neq \emptyset$ for $\alpha, \beta \in A'$ then the change of chart map $\phi_\alpha \circ \overleftarrow{\phi_\beta}$ has a Jacobian matrix with positive determinant.

Example 5

Any manifold admitting a two-chart atlas is orientable if the intersection of the chart domains is connected. For the Jacobian, representing a diffeomorphism between subsets of $\mathbb{R}^n$, cannot have zero determinant. It follows that all of the spheres S^n are orientable because each admits a two-chart atlas : stereographic projection from its north and south poles (cf. Porteous [78] p. 169).

Given a principal fibre bundle (P,G,M) and a manifold F on which G acts on the left, then the fibre bundle associated to (P,G,M) with fibre F is a manifold $(P\times F)/G$ defined by

(i) the right action of G on $P\times F$ is

$$P\times F\times G \longrightarrow P\times F : (u,a,g) \longmapsto (R_g(u), L_{g^{-1}}(a)) ,$$

(ii) the projection map is

$$\Pi_F : (P\times F)/G \longrightarrow M : R_G(u,a) \longmapsto \Pi_P(u)$$

and it is required to be smooth,

(iii) every $x\in M$ has a neighbourhood U such that $\overleftarrow{\Pi_F} U$ is diffeomorphic to $U\times F$.

Example 6

Take $F = G$ and use left translation.

Example 7

The tangent bundle TM is the bundle associated to LM with fibre $\mathbb{R}^n$, where $n = \dim M$. We can identify each fibre $\overleftarrow{\Pi_{\mathbb{R}}}(x)$ with the tangent space $T_xM = \overleftarrow{\Pi_T}(x)$ (cf. §2.1). In particular for S^1 with $F = \mathbb{R}^1 = \mathbb{R}$, $G = \mathbb{R}^*$ we have

$$(LS^1\times\mathbb{R})\times\mathbb{R}^* \longrightarrow LS^1\times\mathbb{R}$$

$$: (x,b,a,g) \longmapsto (x,bg,g^{-1}a) ,$$

for the right action of $\mathbb{R}^*$ on $LS^1\times\mathbb{R}$. Hence the orbit of $\mathbb{R}^*$ through $(x,b,a)\in LS^1\times\mathbb{R}$ is

$$[(x,b,a)] = \{(x,bg,g^{-1}a) \mid g\in\mathbb{R}^*\}$$

which we identify with $(x,a)\in T_xS^1$ because bg runs through all bases for T_xS^1 while g runs through $\mathbb{R}^*$.

The tangent bundle is a special case of the tensor bundles T^k_hM for $k,h = 0,1,\ldots$. These are associated to the frame bundle LM but here F is a tensor product of k copies of $\mathbb{R}^n$ and h copies of its dual $\mathbb{R}^{n*}$. The general linear group acts independently on the factors of the tensor product, as for TM on $\mathbb{R}^n$ and via transposed inverses on $\mathbb{R}^{n*}$ (the contravariant nature of dual taking, again). We identify T^o_oM with $M\times\mathbb{R}$, T^1_oM with

TM, and T^0_1M with TM^* the cotangent vector bundle with fibre $\mathbb{R}^{n*}$. When necessary we shall denote the projection maps onto M by

$$\Pi^k_h : T^k_h M \longrightarrow M .$$

To be more categorical we could introduce a family of functors from Man to V bun , with images these tensor bundles. Such a procedure is outlined in Lang [60] p. 57.

From the tensor bundles we can define the tensor fields, essential in the formulation of physical theories and in the geometrical study of manifolds.

A tensor field of type $\binom{k}{h}$ is a (smooth) section of Π^k_h (cf. II §1.10). That is some

$$W : M \longrightarrow T^k_h M \qquad \text{(in Man)}$$

such that $\Pi^k_h \circ W = I_M$.

The set of such fields forms a vector space, $\mathcal{T}^k_h M$, with pointwise addition and in particular there always exists the zero section field of any type.

A morphism of principle fibre bundles is defined analogously to vector bundle morphisms (cf. §2.1); instead of linear maps on the fibres we need group homomorphisms. A subset (P',G',M) of (P,G,M) is a sub-bundle if the inclusions form a principle fibre bundle morphism; in this case we call (P',G',M) a reduced bundle and we say that G is reducible to G' .

2.3.1 Theorem

The structure group G of a principal fibre bundle (P,G,M) is reducible to a Lie subgroup G' if and only if there is an open cover $\{U_\alpha | \alpha \in K\}$ of M and a set of maps (transition functions)

$$\psi_{\beta\alpha} : U_\alpha \cap U_\beta \longrightarrow G'$$

satisfying for all $x \in U_\alpha \cap U_\beta \cap U_\gamma$

$$\psi_{\beta\alpha}(x) = \psi_{\gamma\beta}(x) . \psi_{\beta\alpha}(x) .$$

Note that G' need not be closed in G (cf. Theorem 2 below).

Proof

Kobayashi and Nomizu [57] p. 51-54. □

Remark

The locally trivial structure of a principal fibre bundle (P,G,M) always provides transition functions taking values in G. For there is an open cover $\{U_\alpha | \alpha\epsilon K\}$ of M with diffeomorphisms

$$\psi_\alpha : \overleftarrow{\Pi}_P U_\alpha \longrightarrow U_\alpha \times G : u \longmapsto (\Pi_P(u), \phi_\alpha(u))$$

satisfying $\phi_\alpha \circ R_g(u) = R_g \circ \phi_\alpha(u)$.

Hence if $u\epsilon\overleftarrow{\Pi}_P (U_\alpha \cap U_\beta)$

$$\left(\phi_\beta \circ R_g(u)\right)\left(\phi_\alpha \circ R_g(u)\right)^{-1} = \phi_\beta(u)\left(\phi_\alpha(u)\right)^{-1} ,$$

depending on $\Pi_P(u)$ only, not on u .

The required functions are given by

$$\psi_{\beta\alpha} \circ \Pi_P(u) = \phi_\beta(u)\left(\phi_\alpha(u)\right)^{-1} .$$

Kobayashi and Nomizu [57] p. 52 show how a principle fibre bundle can be constructed over M , given some transition functions taking values in a Lie group G . We give an example of the construction of a reduced bundle of LS^1, with structure group $\mathbb{Z}$ in §2.6 below.

2.3.2 Theorem

The structure group G of a principal fibre bundle (P,G,M) is reducible to a closed subgroup G' if and only if the associated bundle with fibre G/G' admits a section

$$\sigma : M \longrightarrow E = P/G' .$$

Proof

Kobayashi and Nomizu [57] p. 57. □ (Cf. §2.8 Theorem 5.)

2.3.3 Theorem

A principal fibre bundle (P,G,M) is trivial, that is $P = M \times G$, if and only if P admits a section $\sigma : M \longrightarrow P$.

Proof

Steenrod [96] p. 36. (Cf. also the remark on flat connections at the end of §2.6.) □

Corollary

(P,G,M) is trivial if and only if it is isomorphic to the induced bundle f*Q where Q is the product bundle over a point q and f is the unique constant map.

Proof

The induced bundle, or pullback over M (cf. §2.1) of a constant map $f : M \longrightarrow \{q\}$ is (f^*Q,G,M) where

$$f^*Q = \{(x,u)\in M\times P \mid f(x) = \Pi_P(u)\}$$
$$= \{(x,u)\in M\times \overleftarrow{\Pi}_P(q)\}$$

and the canonical projection is

$$(x,u) \longmapsto x \,.$$

Induced bundles of trivial bundles are trivial and the existence of the above isomorphism is equivalent to the existence of a section $\sigma : M \longrightarrow P$. Details are given in Husemoller [52] p. 48, together with further applications of induced bundles. □

Remark

A <u>sufficient</u> condition for (P,G,M) to be trivial is for M (<u>or</u> the fibres of P) to be contractible (cf. III §1.7). In particular this will happen if M is homeomorphic to $\mathbb{R}^n$ (<u>or</u> if each $\overleftarrow{\Pi}_P(x)$ is homeomorphic to $\mathbb{R}^m$, $\forall x\in M$). (Cf. also Theorem 5 in §2.1.)

2.4 Parallelization

We can always find smooth sections of vector bundles, that is vector fields, but this is not in general true for principal fibre bundles. In particular we shall have an interest in the frame bundle LM (cf. §2.3) and the only compact two-manifolds having a <u>continuous</u> section of their frame bundle are the Klein bottle and torus while the only spheres with this property are S^1, S^3 and S^7 (cf. Brickell and Clark [9] §7.3, Schwartz [87] ch. 10 and Husemoller [52] ch. 15). The difference between finding sections of the vector bundle TM and the fibre bundle LM is that we need $n = \dim M$ linearly independent sections of TM to give a section of LM . In the case of S^2 , the Hairy Ball Theorem asserts that there is not even a continuous section of TS^2 that is never zero.

A manifold M is called parallelizable if LM admits a continuous section; if it exists, such a section is called a parallelization of M . (Note another usage in Hirsch [49] p. 88.)

Manifolds always have local parallelizations about any point because a chart (U,ϕ) there provides a smooth choice of basis for tangent spaces about the point. In consequence of this there is always a local diffeomorphism $\overleftarrow{\Pi}_L U \equiv U\times G$. Any manifold with a global chart is inevitably parallelizable; for example $\mathbb{R}^n$ admits the identity chart. If a manifold M is parallelizable then TM is globally homeomorphic to $M\times\mathbb{R}^n$, and if the parallelization is a C^k-section of LM then this homeomorphism is a C^k-diffeomorphism. Thus, LM is trivial if and only if M is parallelizable.

We have seen how the frame and tangent bundles are related functorially by the tangent bundle functor (cf. §2.2) which gives an exact sequence (cf. §2.3)

$$\{\underline{0}\} \longrightarrow \ker(T\circ\Pi_L) \longrightarrow TLM \longrightarrow \Pi_L^*(TM) \longrightarrow \{\underline{0}\} .$$

In the case that M is paracompact this sequence splits and LM is parallelizable (cf. Hirsch [49] p. 98). We shall see later that this is related to the idea of a connection (cf. §2.6).

We collect here some results associated with parallelizations (cf. Brickell and Clark [9], Husemoller [52], Kobayashi and Nomizu [57], Hirsch [49]):

(i) If M is compact and connected then TM admits a nowhere-zero section if and only if the Euler characteristic $\chi(M)$ is zero. Since

$$\chi(S^n) = 1 + (-1)^n ,$$

only odd-dimensional spheres admit global nowhere-zero vector fields. Also, any compact odd-dimensional manifold M has $\chi(M) = 0$, so does any closed orientable manifold with a nowhere-zero tangent vector field.

(ii) Any non-compact paracompact manifold admits a nowhere-zero vector field.

(iii) If M admits a nowhere-zero vector field then so does $M \times N$ for any manifold N.

(iv) If M and N are parallelizable then so is $M \times N$.

(v) Given a parallelization

$$p : M \longrightarrow LM : x \longmapsto (p_i(x))$$

then the set of all smooth maps

$$f : M \longrightarrow M$$

satisfying

$$Df \circ p_i = p_i \circ Df \,, \quad (\forall i = 1,2,\ldots, \dim M)$$

forms a group, <u>the automorphism group of</u> p.

The group of automorphisms of a parallelization for a connected manifold is either a <u>discrete group</u> (orbits are discrete sets of points) or it can be given the structure of a <u>Lie group</u>, with a free action on M (cf. §1.5). These automorphisms are always isometries with respect to the standard metric structure induced by a parallelization (cf. §2.8 below).

(vi) Every parallelizable manifold is orientable.

(vii) Every parallelizable manifold is metrizable and therefore paracompact.

(viii) Every Lie group is parallelizable.

(ix) Any orientable, compact 3-manifold is parallelizable (in fact, the compactness can be relaxed, cf. Steenrod [96] p.203 and p.221.)

<u>Example</u>

A non-trivial parallelization of $\mathbb{R}^2$ is given by

$$p : \mathbb{R}^2 \longrightarrow L\mathbb{R}^2 : (x,y) \longmapsto (e^x\partial_1, e^x\partial_2).$$

In §2.8 it will be used to illustrate that a parallelization always determines a metric structure and a connection (cf. §2.6) that is compatible with this structure (sizes of vectors are unaltered by parallel transport) but the connection need not be symmetric.

2.5 *Lie algebra*

A Lie algebra is a vector space ξ together with a product structure

$$\xi\times\xi \longrightarrow \xi : (\alpha,\beta) \longmapsto [\alpha,\beta]$$

satisfying

(i) bilinearity

(ii) $(\forall\alpha\in\xi)\ [\alpha,\alpha] = 0$

(iii) the Jacobi identity $(\forall\alpha,\beta,\gamma\in\xi)$

$$[[\alpha,\beta],\gamma] + [[\beta,\gamma],\alpha] + [[\gamma,\alpha],\beta] = 0 .$$

The second part of Hochschild [50] is devoted to a development of the theory of Lie algebras. We shall be concerned with algebras over $\mathbb{R}$, but complex numbers could equally well be used and the complex analytic Lie groups and their algebras are also studied in [50] ch. 17. In fact, it is the natural generation of Lie algebras from Lie groups that makes them so important:

2.5.1 Theorem

If ξ is a finite-dimensional Lie algebra over $\mathbb{R}$ then there exists a connected Lie group G whose Lie algebra is isomorphic to ξ.

Proof

Hochschild [50] ch. 12. □

Now we see how the Lie algebra arises :

A left-invariant vector field of a Lie group G is any $w : G \longrightarrow TG$ satisfying

$$(\forall g,h\in G)\ DL_g \circ w(h) = w\circ L_g(h) = w(gh) ,$$

that is, fixed under differentials of left translations. Such fields form a vector space and in fact a Lie algebra, ξ, under the Lie bracket composition

$$[w,v](f) = w(v(f)) - v(w(f))$$

$$\forall w,v \in \xi, \ \forall C^{\infty} f : G \longrightarrow \mathbb{R} \qquad \text{(cf. §2.2)}.$$

This composition is of course available to all vector fields on G, i.e. to T^1G, and the inclusion

$$\xi \hookrightarrow T^1G$$

is a Lie algebra homomorphism.

We can visualise ξ through an isomorphism (in Vec) with the tangent space to the identity:

$$\xi \equiv T_eG : \gamma \longmapsto \gamma(e)$$

so we see $\dim \xi = \dim G$.

Example 1

In the case $G = G\ell(n;\mathbb{R})$ the tangent space T_aG at any $a \in G$ is the matrix space

$$T_aG = \{(a,A) \mid A \text{ is an } n\times n \text{ real matrix}\}$$

so $T_aG \equiv \mathbb{R}^{n^2}$.

The Lie algebra $\xi = \xi\ell(n;\mathbb{R})$ consists essentially of all $n\times n$ real matrices A , which determine fields by

$$A : G \longrightarrow TG : a \longmapsto (a,A)$$

and the Lie bracket is the composite matrix product:

$$[A,B] = AB - BA .$$

We illustrate the left-invariance property for the case $n = 1$. Then $G = \mathbb{R}^*$, $\xi = \mathbb{R}$ and

$$(\forall \gamma \in \mathbb{R}) \ \gamma : \mathbb{R}^* \longrightarrow T\mathbb{R}^* : a \longmapsto (a, \gamma_a)$$

is left-invariant if

$$(\forall a,g \in \mathbb{R}^*) \ DL_g\gamma(a) = \gamma L_g(a) = \gamma(ga) .$$

Hence at the identity $e = 1 \in \mathbb{R}^*$

$$g\gamma(e) = \gamma(ge) = \gamma(g) ,$$

so we require

$$\gamma(a) = a\gamma(1) = a\gamma_1 , \text{ say} .$$

The adjoint representation Ad of a Lie group G in its Lie algebra ξ is obtained from the derivative of the automorphism

$$ad(g) = L_g R_{g^{-1}} : G \longrightarrow G : h \longmapsto ghg^{-1} \quad (\forall g \in G)$$

namely,

$$Ad(g) : \xi \longrightarrow \xi : \gamma \longmapsto D(L_g R_{g^{-1}})\gamma = Dad(g)\gamma .$$

It follows that left-invariance is preserved under each map Ad(g). For more theory see Hochschild [50] and Serre [88], and for applications in physics see DeWitt and DeWitt [23].

Example 2

The adjoint representation is trivial for any $g \in G\ell(1;\mathbb{R})$; because

$$L_h R_{g^{-1}} : \mathbb{R}^* \longrightarrow \mathbb{R}^* : h \longmapsto ghg^{-1} = h$$

implies $Ad(g) = I_{\mathbb{R}}$.

The following is basic to local geometry of Lie groups:

2.5.2 Theorem

Every $\gamma \in \xi$ generates a one-parameter subgroup of G .

Proof (Outline only)

Let $c_\gamma : t \longmapsto \gamma_t$ be the integral curve in G (cf. §2.2), determined for $|t| < \varepsilon$ for some real $\varepsilon > 0$, with initial conditions

$$\gamma_o = c_\gamma(0) = e \in G$$

$$\dot{c}_\gamma(0) = \gamma(e) \in T_e G \equiv \xi \quad \text{(as a vector space)} .$$

We define

$$\phi_t : G \longrightarrow G : g \longmapsto L_g(\gamma_t), \quad |t| < \varepsilon .$$

Since this is well-defined for all $g \in G$ it admits an extension to all $t \in \mathbb{R}$ and we have a group because where both sides are defined we find

$$L_g(\gamma_{t+s}) = L_g(\gamma_t) \circ L_g(\gamma_s) .$$

The required subgroup of G is

$$G_\gamma = \{\phi_t(e) \mid t \in \mathbb{R}\} . \qquad \square$$

Next we obtain the exponential map

$$\exp : \xi \longrightarrow G : \gamma \longmapsto \gamma_1$$

that is, each field is followed along its integral curve to unit distance from the identity (cf. Hochschild [50] p. 79 and Dodson and Poston [28] p. 354.)

Example 3

In the case $G = G\ell(n;\mathbb{R})$ the exponential map coincides with the usual function for matrices γ

$$\exp \gamma = \sum_{k=0}^{\infty} \gamma^k/k! \quad .$$

Something of why this is so can be gleaned from the situation when $n = 1$. Take $\gamma \in \xi\ell(1;\mathbb{R}) = \mathbb{R}$ and seek the integral curve $c_\gamma : (-\varepsilon,\varepsilon) \longrightarrow \mathbb{R}^*$ for which

$$c_\gamma(0) = e = 1$$

$$\dot{c}_\gamma(0) = \gamma(1) = \gamma_o$$

$$\dot{c}_\gamma(t) = \gamma \circ c_\gamma(t) .$$

Evidently, we are seeking a particular solution of the differential equation

$$\frac{d}{dt} c_\gamma = \gamma_o \, c_\gamma ,$$

and it is actually

$$c_\gamma : (-\varepsilon,\varepsilon) \longrightarrow \mathbb{R}^* : t \longmapsto e^{\gamma_o t} .$$

Hence the one-parameter subgroup is given by maps of the form

$$\phi_t : \mathbb{R}^* \longrightarrow \mathbb{R}^* : g \longmapsto g\, e^{\gamma_o t}$$

and the exponential map is

$$\exp : \mathbb{R} \longrightarrow \mathbb{R}^* : \gamma \longmapsto e^\gamma .$$

We defined in §1.5 the action of a Lie group G on a manifold; now we can bring in the role of its Lie algebra. Suppose then that G acts on the right of a manifold P

$$P\times G \longrightarrow P : (u,g) \longmapsto R_g(u) .$$

From above we find for each $\gamma\in\xi$ a one-parameter subgroup of G

$$G = \{\gamma_t(e) = \exp t\gamma \,|\, t\in\mathbb{R}\} ,$$

acting on the right of P by inclusion in G. The following properties are readily checked :

(i) through each $u\in P$ there is a smooth curve

$$\gamma_u : t \longmapsto R_{\gamma_t(e)}(u)$$

with tangent vector $\dot{\gamma}_u(0)\in T_uP$ at $t = 0$.

(ii) there is a map

$$\Phi : \xi \longrightarrow T^1P : \gamma \longmapsto \dot{\gamma}$$

where T^1P is the Lie algebra of all vector fields on P and

$$\dot{\gamma} : P \longrightarrow TP : u \longmapsto \dot{\gamma}_u \quad \text{(as in (i))}$$

is called the <u>fundamental vector field corresponding to</u> $\gamma\in\xi$. Then Φ is a Lie algebra homomorphism and if the action of G is free (only e has a fixed point) then $\dot{\gamma}_u$ is not the zero vector at any $u\in P$, for non-zero $\gamma\in\xi$.

<u>Example 4</u>

Consider $G = O(2;\mathbb{R})$, the orthogonal subgroup of $G\ell(2;\mathbb{R})$. Then for any

$$\gamma = \begin{bmatrix} 0 & \theta \\ \theta & 0 \end{bmatrix} \in \xi$$

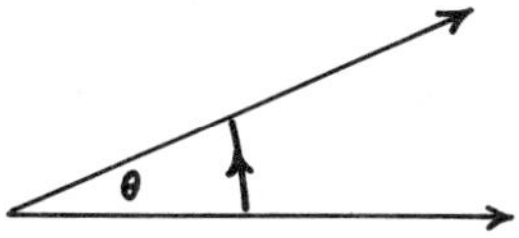

we have

$$\exp\gamma = \begin{bmatrix} \cos\theta & \sin\theta \\ -\sin\theta & \cos\theta \end{bmatrix} .$$

<u>Example 5</u>

For $G = G\ell(1;\mathbb{R}) = \mathbb{R}^*$ and any $\gamma\in\xi\ell(1;\mathbb{R}) = \mathbb{R}$,

$$G_\gamma = \{e^{t\gamma} = \exp t\gamma \mid t\in\mathbb{R}\} \subset \mathbb{R}^* .$$

Example 6

Using the action of $\mathbb{R}^*$ on $\mathbb{R}^2$ given by

$$\mathbb{R}^2\times\mathbb{R}^* \longrightarrow \mathbb{R}^2 : ((x,y),g) \longmapsto (gx,gy)$$

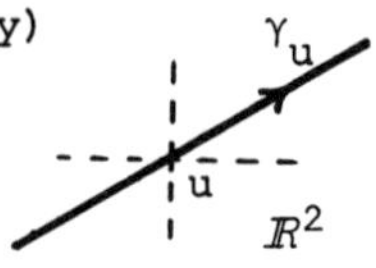

we obtain a curve in $\mathbb{R}^2$ through $u = (x,y)$:

$$\gamma_u : t \longmapsto R_{e^{t\gamma}}(u) = (xe^{t\gamma}, ye^{t\gamma}) .$$

The tangent vector to this at any t is

$$\dot\gamma_u(t) = \left((xe^{t\gamma}, ye^{t\gamma}), (\gamma xe^{t\gamma}, \gamma ye^{t\gamma})\right) \in T_{\gamma_u(t)}\mathbb{R}^2 .$$

Evaluating this at $t = 0$ yields the map

$$\gamma : \mathbb{R}^2 \longrightarrow T\mathbb{R}^2 : u \longmapsto (u, \gamma u)$$

and in this example $\dot\gamma_u = \underline{0}$ at $u = \underline{0}$ because all $\gamma\in\mathbb{R}$ have a fixed point there.

Let (P,G,M) be a principal fibre bundle. Then for all $\gamma\in\xi$ we note the following properties of the fundamental vector field $\dot\gamma$:

(i) G translates each fibre along itself freely, so the linear map

$$\xi \longrightarrow T_uP : \gamma \longmapsto \dot\gamma_u$$

has trivial kernel and is injective for all $u\in P$.

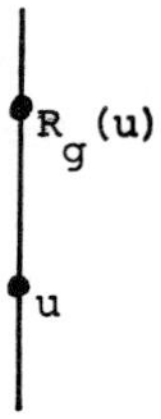

(ii) We can obtain $\dot\gamma$ by means of $(\forall u\in P)$

$$\sigma_u : G \longrightarrow P : g \longmapsto R_g(u)$$

for $D\sigma_u : TG \longrightarrow TP$ restricts to give at the identity

$$T_eG \longrightarrow T_uP : \gamma \longmapsto \dot\gamma_u$$

where as usual we identify T_eG with ξ .

(iii) Given $\Pi_P(u) = x$

$$D\Pi_P\,\dot\gamma_u = \underline{0} \in T_xM ,$$

i.e. $\dot\gamma_u$ is tangent to the fibre containing u .

(iv) As a vector space ξ is isomorphic to the vertical subspace G_u over any $u \in P$ by the map

$$\xi \longrightarrow G_u : \gamma \longmapsto \dot{\gamma}_u .$$

Hence $\dim G_u = \dim \xi = \dim G$.

Example 7

Take the case of a frame bundle LM with $G = G\ell(n;\mathbb{R})$, $\xi = \xi\ell(n;\mathbb{R})$. We display the above maps in component form (cf. §2.3 Example 3) for any $u = (x^i, b^i_j) \in LM$.

$$\sigma_u : G \longrightarrow LM : (g^i_j) \longmapsto (x^i, g^i_m b^m_j)$$

$$D\sigma_u : T_e G \longrightarrow T_u LM : (\delta^i_j, b^i_j) \longmapsto (x^i, b^i_j, 0, \gamma^i_m b^m_j)$$

$$\xi \longrightarrow T_u LM : (\gamma^i_j) \longmapsto (\gamma^i_m b^m_j)$$

$$(\dot{\gamma}^i_j) : LM \longrightarrow TLM : u \longmapsto (x^i, b^i_j, 0, \gamma^i_m b^m_j) .$$

Note the use of the summation convention for matrix products

$$[g^i_m b^m_j] = [g^i_m] \; [b^m_j] .$$

In particular, for $LS^1 = S^1 \times \mathbb{R}^*$ we have $\gamma \in \xi\ell(1;\mathbb{R}) = \mathbb{R}$ and $u = (x,b)$ so we find :

$$\dot{\gamma} : LS^1 \longrightarrow TLS^1 : (x,b) \longmapsto (x,b,0,\gamma b)$$

$$\sigma_u : \mathbb{R}^* \longrightarrow LS^1 : g \longmapsto (x,gb)$$

$$D\sigma_u : T_1\mathbb{R}^* \longrightarrow T_u LS^1 : (1,\gamma) \longmapsto (x,b,0,\gamma b) \text{ at } g=1$$

$$T_1\mathbb{R}^* \equiv \mathbb{R} : (1,\gamma) \longmapsto \gamma .$$

A Lie algebra ξ is called <u>nilpotent</u> if there is a positive integer n such that for every n-tuple $(\alpha_1, \alpha_2, \ldots, \alpha_n)$ of elements from ξ we have

$$\Delta_{\alpha_1} \cdot \Delta_{\alpha_2} \cdot \cdot \cdot \Delta_{\alpha_n} = 0$$

where

$$\Delta_\alpha : \xi \longrightarrow \xi : \beta \longmapsto (\alpha\beta - \beta\alpha) .$$

2.5.3 Theorem

Let G be a connected Lie group with nilpotent Lie algebra ξ .

Then the exponential map is surjective.

If G is also simply-connected then the exponential map is an isomorphism in Man.

Proof

Hochschild [50] p. 136. □

2.6 *Connection; parallel transport, geodesic, holonomy group*

The early work of Weyl and others on connections (in tangent bundles) was much stimulated by the growth of interest in general relativity (cf. Weyl [101] §14 and Notes 9, 10). More recently the pure geometric formulation, notably by Ehresmann [32], extended the notion to general bundles. Interestingly, these generalizations have proved very useful in current particle physics under the name of Yang-Mills field theory (cf. Hermann [45]). There are three helpful ways to view a connection in a vector bundle (M,E,Π_E) :

(i) as a 'covariant derivative' of sections of the bundle E , with respect to tangent vector fields;

(ii) as a specification of which vector in $\overleftarrow{\Pi}_E \circ c(t)$ is 'parallel' to a given vector in $\overleftarrow{\Pi}_E \circ c(0)$, with respect to a given curve c in M ;

(iii) as a smooth 'splitting' of the tangent bundle TE into 'horizontal' and 'vertical' subspaces over each $u \in E$. A modern geometric motivation for the definition of a connection is given in Dodson and Poston [28] Ch. 8, with pictorial representation of the above three roles. Our plan for the present section is to define a connection in a principal fibre bundle then deduce various properties, including its influence on associated vector bundles. We shall mainly follow Kobayashi and Nomizu [57] for definitions, taking some examples from Dodson [26].

A connection ∇ in a principle fibre bundle (P,G,M) is an assignment of a subspace H_u of T_uP for all $u \in P$ such that

(i) $T_u P = H_u \oplus G_u$

$: X \longmapsto X_H \oplus X_G$ smoothly on TP, where G_u is the vertical subspace (cf. §2.3) ;

(ii) $DR_g H_u = H_{R_g(u)}$ $\forall g \in G$.

We call H_u the <u>horizontal subspace</u> at $u \in P$. The derivative of Π_P induces an isomorphism

$$H_u \equiv T_{\Pi_P(u)} M \quad (\forall u \in P)$$

and we already have a composite isomorphism (cf. §2.5)

$$G_u \equiv \xi \equiv T_e G \quad (\forall u \in P)$$

<u>Example 1</u>

Consider $P = LS^1$; a <u>constant</u> connection in LS^1 is given by any $\lambda \in \mathbb{R}$ if we put

$$\left(\forall u = (x,b) \in LS^1\right) \; H_{(x,b)} = \{(x,b,p,-\lambda bp) \mid p \in \mathbb{R}\} \; .$$

Then the smooth decomposition of $T_u LS^1$ is $T_u LS^1 = H_u \oplus \mathbb{R}^*_u$:

$$(x,b,p,q) = (x,b,p,-\lambda bp) \oplus (x,b,0,q+\lambda bp) \; .$$

We check the compatibility with the right action of $g \in \mathbb{R}^*$,

$$R_g : LS^1 \longrightarrow LS^1 : (x,b) \longmapsto (x,bg)$$

$$DR_g : TLS^1 \longrightarrow TLS^1 : (x,b,p,q) \longmapsto (x,bg,p,bg)$$

$$DR_g H_{(x,b)} = H_{(x,bg)} \; .$$

Also, the differential of the projection Π_L is

$$D\Pi_L : TLS^1 \longrightarrow TS^1 : (x,b,p,q) \longmapsto (x,p)$$

and it gives the isomorphism

$$H_{(s,b)} \equiv T_x S^1 : (x,b,p,-\lambda bp) \longmapsto (x,p) \; .$$

Of course, $H_{(x,b)}$ only <u>looks</u> horizontal in the standard embedding $LS^1 \hookrightarrow \mathbb{R}^3$ if $\lambda = 0$, which would be the connection giving the usual parallelism structure.

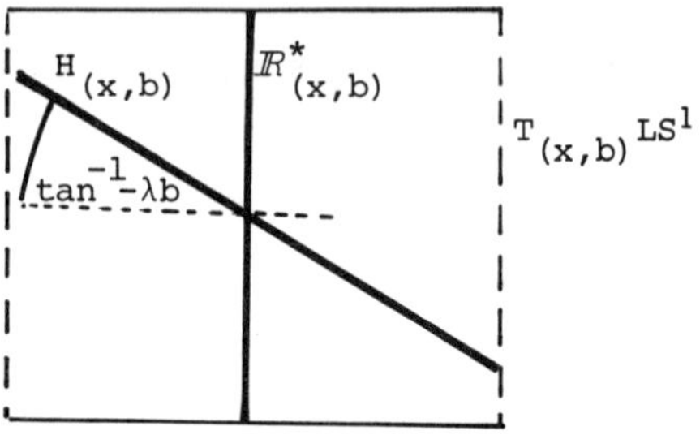

The <u>connection form</u> of a connection ∇ in (P,G,M) is the smooth map

$$\omega : TP \longrightarrow \xi : X_H \oplus X_G \longmapsto \gamma$$

$$\text{with } X_G(u) = \dot{\gamma}_u \quad (\text{cf. §2.5}).$$

<u>Properties of the connection form</u> (cf. Kobayashi and Nomizu [57] p. 64 for proofs.)

(i) $(\forall \gamma \in \xi) \ \omega(\dot{\gamma}_u) = \gamma$.

(ii) $\omega(X) = \underline{0} \iff X = X_H \oplus \underline{0}$.

(iii) $(\forall g \in G, \ \forall X \in T^1P)$

$$\omega \circ DR_g X = Ad(g^{-1})\omega(X), \quad (\text{cf. §2.5}) .$$

(iv) Connections and connection forms determine one another uniquely. □

<u>Example 2</u>

We continue the preceding example:

$$(x,b,p,-\lambda bp) \oplus (x,b,0,q+\lambda bp) = (x,b,p,q) ;$$

since $\dot{\gamma}(x,b) = (x,b,0,\gamma b)$ (cf. §2.5 Ex. 7) ,

$\omega(x,b,p,q) = \gamma = (q+\lambda bp)/b \in \xi\ell(1;\mathbb{R}) = \mathbb{R}$.

We can see the above properties for this example:

(i) $\omega \circ \dot{\gamma}(x,b) = \omega(x,b,0,\gamma b) = \gamma b/b = \gamma , (\forall \gamma \in \mathbb{R})$;

(ii) $(q+\lambda bp)/b = 0 \iff q+\lambda bp = 0$.

(iii) Given $X : LS^1 \longrightarrow TLS^1 : (x,b) \longmapsto (x,b,p,q)$, and $g \in \mathbb{R}^*$, $Ad(g^{-1})$ is the identity (cf. §2.5) :

$$Ad(g^{-1})\omega(X) = D(L_{g^{-1}}R_g)\omega(X) = \omega(X) .$$

Next,

$$DR_g : TLS^1 \longrightarrow TLS^1 : (x,b,p,q) \longmapsto (x,bg,p,qg)$$

and

$$\omega(X) : \mathbb{R}^* \longrightarrow T\mathbb{R}^* : a \longmapsto \big(a,a(q+\lambda bp)/b\big)$$

so, as required,

$$\omega \circ DR_g(X) : \mathbb{R}^* \longrightarrow T\mathbb{R}^*$$

satisfies property (iii) :

$$\begin{aligned} \omega \circ DR_g(x,b,p,q) &= \omega(x,bg,p,qg) \\ &= (qg+\lambda bgp)/bg = (q+\lambda bp)/b. \end{aligned}$$

(iv) Evidently ω and λ determine one another uniquely. Next we come to an important construction process:

2.6.1 Theorem

Given a connection ∇ in a principal fibre bundle (P,G,M), every vector field $w : M \longrightarrow TM$ has a unique horizontal lift

$$w^\dagger : P \longrightarrow TP : u \longmapsto w^\dagger(u)$$

with the following properties.

(i) $D\Pi_P w^\dagger(u) = w \circ \Pi_P(u)$ and $w^\dagger(u) \in H_u$.

(ii) $w^\dagger$ is invariant by R_g $(\forall g \in G)$, and every horizontal vector field on P invariant by G is the horizontal lift of some vector field on M .

(iii) $w^\dagger + v^\dagger = (w+v)^\dagger$.

(iv) If $f \in T^0M$ we put $f^\dagger = f \circ \Pi_P$

then for any vector field $w : M \longrightarrow TM$

$$(f.w)^\dagger = f^\dagger . w^\dagger : P \longrightarrow TP .$$

(v) $[v^{\uparrow}, w^{\uparrow}]_H = [v,w]^{\uparrow}$ (H denotes horizontal component).

Proof

Kobayashi and Nomizu [57] p. 65. □

Corollary

Every piecewise C^1- curve $c : [0,1) \longrightarrow M$ has a unique *horizontal lift* curve

$$c^{\uparrow} : [0,1) \longrightarrow P, \text{ through } u_0 \in \overleftarrow{\Pi}_P \circ c(0) \;,$$

with $\Pi_P \circ c^{\uparrow} = c \;, \quad c^{\uparrow}(0) = u_0 \;;$

and the tangent vector field of $c^{\uparrow}$ is the horizontal lift of $\dot{c}$, that is $\dot{c}^{\uparrow}$.

Moreover, the map

$$\tau_t : \overleftarrow{\Pi}_P \circ c(0) \longrightarrow \overleftarrow{\Pi}_P \circ c(t) : u_0 \longmapsto c^{\uparrow}(t) \;,$$

called *parallel transport along* c, commutes with the action of G on P and is a *diffeomorphism*.

Proof

Kobayashi and Nomizu [57] p. 68-70, cf. also Dodson and Poston [28] Ch. 8 for more discussion. □

We shall look at examples of these horizontal lifts shortly but first it is convenient to interpret the connection in (P,G,M) as a derivation in an associated vector bundle (cf. §2.1 and §2.3). For this purpose we must extend the notion of parallel transport in a principal fibre bundle to its associated vector bundles. (Note that it is not difficult to make any vector bundle an associated bundle of some principal bundle.)

Let ∇ be a connection in (P,G,M) and let (M,E,Π_E) be a vector bundle with fibre V associated to (P,G,M). We effect a splitting of each tangent space T_aE for $a \in E$ into a direct sum

$$T_aE = H_a \oplus G_a \;,$$

with $G_a = \{X \in T_aE \mid D\Pi_E X = 0 \in T_{\Pi_E(a)}M\}$

as a *vertical subspace*, and H_a the *horizontal subspace* defined as follows (cf. §2.3). Denote the canonical factorisation by G

of $P \times V$ by:

$$f_G : P \times V \longrightarrow (P \times V)/G = E$$

and choose any $(u,y) \in \overleftarrow{f_G}(a)$, that is $[(u,y)]_G = a$.

Fix y and introduce the map

$$f : P \longrightarrow E : v \longmapsto f_G(v,y) .$$

The derivative of this map at $u \in P$ takes us to $T_a E$

$$D_u f : T_u P \longrightarrow T_a E$$

and we define the horizontal subspace of $T_a E$ by

$$H_a = D_u f(H_u) .$$

This construction makes H_a independent of the choice of (u,y) and does indeed yield the required direct sum with G_a ; it varies smoothly with a because G_a varies smoothly, and so does H_u with u . Now we can obtain unique horizontal lifts from M to E in much the same way as we did from M to P . Hence, we obtain also parallel transport between fibres of E along a given curve c in M :

$$\tilde{\tau}_t : \overleftarrow{\Pi_E} \circ c(0) \longrightarrow \overleftarrow{\Pi_E} \circ c(t)$$

$$: a_o \longmapsto f_G(c^{\dagger}(t),y)$$

where y is any element of V with $a_o = [(u_o,y)]_G$, and $c^{\dagger}$ is the horizontal lift of c to P . As before this map is a diffeomorphism and since here the domain and image are vector spaces it is an <u>isomorphism</u>.

We say that a section v of E defined on an open set U of M is <u>parallel along a given curve</u> $c : [0,1] \longrightarrow U$, if we have

$$\tilde{\tau}_t \circ v \circ c(0) = v \circ c(t) ;$$

if v is parallel along <u>all</u> curves in U we say that v is a <u>parallel section</u>.

Let (P,G,M) be a principal fibre bundle with connection ∇ and let (M,E,Π_E) be a vector bundle with fibre V associated to (P,G,M). Given any C^1- curve $c : [0,1) \longrightarrow M$ and a section v of E defined along c , that is

$$\Pi_E \circ v \circ c(t) = c(t) \qquad \forall\, t \in [0,1) ,$$

then the <u>covariant derivative</u> of v along c is

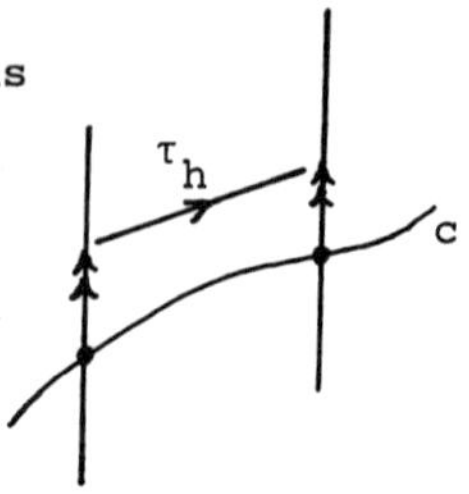

$$\nabla_{\dot{c}(t)} v = \lim_{h \to 0} \frac{1}{h}\left(\overleftarrow{\tau}_h v \circ c(t+h) - v \circ c(t)\right)$$

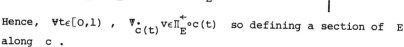

where τ_h is the parallel transport isomorphism

$$\tau_h : \overleftarrow{\Pi}_E \circ c(t) \longrightarrow \overleftarrow{\Pi}_E \circ c(t+h) .$$

Hence, $\forall t \in [0,1)$, $\nabla_{\dot{c}(t)} v \in \overleftarrow{\Pi}_E \circ c(t)$ so defining a section of E along c .

<u>Properties of the covariant derivative</u>

(i) v is parallel along c $\iff \nabla_{\dot{c}(t)} v = \underline{0}\ \forall t$.

(ii) $\nabla_{\dot{c}(t)} (v+w) = \nabla_{\dot{c}(t)} v + \nabla_{\dot{c}(t)} w$.

(iii) If f is a real function defined along c then

$$\nabla_{\dot{c}(t)} (f\,v) = f \circ c(t) . \nabla_{\dot{c}(t)} v + \dot{c}(t)(f) . v \circ c(t) .$$

(iv) We can extend the application to $\mathcal{E}_U$, sections of E defined on some open neighbourhood U of $x \in M$, hence obtaining a map

$$\nabla_x : T_x M \times \mathcal{E}_U \longrightarrow \mathcal{E}_U$$

$$: (r,v) \longmapsto \nabla_{\dot{c}(0)} v = \nabla_r v, \text{ say} ,$$

where c is any curve in M with $\dot{c}(0) = r$.

(In particular, we note the important case when $\mathcal{E}_U$ is one of the spaces of tensor fields $T^k_h M$, Example 3 below.)
Using this notation we further obtain:

(v) $\nabla_{r+s} v = \nabla_r v + \nabla_s v$.

(vi) $\nabla_r (v+w) = \nabla_r v + \nabla_r w$.

(vii) $\nabla_{\lambda r} v = \lambda \nabla_r v \qquad (\forall \lambda \in \mathbb{R})$.

(viii) $\nabla_r f v = f(x) \nabla_r v + r(f) . v(x)$, $\qquad (\forall f : U \to \mathbb{R})$.

(ix) If r and s are vector fields on M, i.e. $r, s \in TM$, and v, w are sections of E on M then properties (v) - (viii) remain valid but the expressions are now sections of E on M.

Proof

Kobayashi and Nomizu [57] Ch. 3, Dodson and Poston [28] Ch. 8. □

Example 3

In the particular case of a connection ∇ in a frame bundle the induced covariant derivation acts on tensor fields on M. We display the action in local co-ordinates for tangent vector fields. Suppose that (U,ϕ) is a chart on M, then we have a local field of frames

$$(\partial_i) : U \longrightarrow LM : x \longmapsto (\partial_i)_x$$

with respect to which any field $v : M \longrightarrow TM$ has a local expression

$$v(x) = v^i(x)\partial_i(x) \qquad \forall x \in U .$$

We can safely omit reference to the point x and write $v = v^i \partial_i \in T^1 U$. Now, the covariant derivation is linear and so it is uniquely determined locally in U by its action on the basis fields (∂_i) :

$$\nabla_{\partial_i} \partial_j = \Gamma^k_{ij} \partial_k \qquad \forall\ i,j .$$

The arrays of real functions Γ^k_{ij} are the Christoffel symbols, or connection components for ∇ with respect to the chart (U,ϕ).

We use these Γ^k_{ij} to display how ∇ splits each $T_u LM$. Firstly we note that $u \in LM$ admits a local expression: $u = (x^i, b^i_j) \in \mathbb{R}^n \times \mathbb{R}^{n^2}$ where $\Pi_L(u) = x \in U$ and $\phi(x) = (x^i) \in \mathbb{R}^n$ and the frame determined by u for $T_x M$ is

$$b^i_1 \partial_i,\ b^i_2 \partial_i,\ \ldots,\ b^i_n \partial_i .$$

Now, given $Y = (x^i, b^i_j, X^i, B^i_j) \in T_u LM$

we have (cf. Examples 1,2 for the case n = 1):

$$Y = Y_H \oplus Y_G = (x^i, b^i_j, X^i, -b^k_j \Gamma^{\ i}_{k\ell} X^\ell) \oplus (x^i, b^i_j, 0, B^i_j + b^k_j \Gamma^{\ i}_{k\ell} X^\ell) .$$

Also, in local co-ordinates, the connection form ω sends :

Y to that matrix $[\gamma^i_j] = \gamma\in\xi\ell(n;\mathbb{R})$ such that:

$$\dot{\gamma}_u = Y_G, \quad \text{that is} \quad [\gamma^i_j] = [B^i_j + b^k_j \Gamma^i_{k\ell} X^\ell][b^i_j]^{-1} .$$

Omitting the indices from these matrices we write

$$\gamma = (B+b\Gamma X)b^{-1} .$$

Now we display the property

$$(\forall g\in G)\,\omega\circ DR_g Y = Ad(g^{-1})\gamma .$$

We have $DR_g(Y)$, for $g = [g^i_j]\in G\ell(n;\mathbb{R})$, given by

$$(x^i, g^i_m b^m_j, X^i, -g^k_m b^m_j \Gamma^i_{k\ell} X^\ell)$$

$$\oplus\ (x^i, g^i_m b^m_j, 0, g^i_m B^m_j + g^k_m b^m_j \Gamma^i_{k\ell} X^\ell) .$$

Then

$$\omega\circ DR_g Y = \eta = [\eta^i_j] \text{ such that}$$

$$(\eta^i_k g^k_m b^m_j) = (g^i_m B^m_j + g^k_m b^m_j \Gamma^i_{k\ell} X^\ell) .$$

Omitting indices from the matrices and solving for η :

$$\begin{aligned}\eta &= (gB+gb\Gamma X)(gb)^{-1} \\ &= g(B+b\Gamma X)b^{-1}g^{-1} \\ &= R_{g^{-1}}\circ L_g(\gamma) = Ad(g^{-1})\gamma .\end{aligned}$$

Next we find the local differential equation for the horizontal lift $c^\uparrow$ of a curve

$$c : [0,1] \longrightarrow U \text{ which locally appears as}$$

$$\phi\circ c : [0,1] \longrightarrow \mathbb{R}^n : t \longmapsto (c^i(t)) ,$$

where (U,ϕ) is some chart on M . We have

$$c^\uparrow : [0,1] \longrightarrow LU , \quad \Pi_L\circ c^\uparrow = c .$$

Locally,

$$\dot{c} = \dot{c}^i \partial_i \quad \text{with} \quad \dot{c}^i = \frac{dc^i}{dt} ,$$

and since each $c^\uparrow(t)$ is a frame at $c(t)$ it can also be expressed via the frame $(\partial_i)_{c(t)}$:

$$c^{\uparrow}(t) = (b^i_j \partial_i)_{c(t)} .$$

Now, for each $j = 1,2,\ldots,n$: $b^i_j \partial_i \in T_{c(t)}M$, we want it to be parallel along c and this means

$$\nabla_c (b^i_j \partial_i) = \underline{0} , \quad j = 1,2,\ldots,n .$$

From the properties of ∇ this equation becomes

$$\nabla_{\dot{c}^i \partial_i} (b^i_j \partial_i) = \left(\frac{d}{dt} b^i_j + b^k_j \dot{c}^\ell \Gamma^i_{k\ell} \right) \partial_i = \underline{0} ,$$

so the differential equations to be solved are

$$\frac{d}{dt} b^i_j + b^k_j \dot{c}^\ell \Gamma^i_{k\ell} = 0 , \quad i,j = 1,2,\ldots,n.$$

These do have a unique solution $b^i_j(t)$ satisfying given initial conditions

$$[b^i_j(0)] = [\beta^i_j]$$

by the classical Cauchy theorem (note that the specification of $c^{\uparrow}(0)$ also fixes $\dot{c}^{\uparrow}(0)$ via the equation). Then we have the parallel transport map along c :

$$\tau_t : \overleftarrow{\Pi}_L \circ c(0) \longrightarrow \overleftarrow{\Pi}_L \circ c(t)$$

$$: (\beta^i_j \partial_i) \longmapsto (b^i_j(t) \partial_i) .$$

Example 4

Further to Examples 1, 2, consider the case of the constant connection $\lambda = \Gamma^1_{11} \in \mathbb{R}$ in the frame bundle LS^1 . The curve (with fixed $\alpha \in \mathbb{R}$)

$$c : [0,1) \longrightarrow S^1 : t \longmapsto \alpha t$$

where we consider S^1 as $\mathbb{R}$ (mod 1) ,
has horizontal lift

$$c^{\uparrow} : [0,1) \longrightarrow LS^1 : t \longmapsto (\alpha t, b(t))$$

with

$$\frac{db}{dt} + b\lambda\alpha = 0 ,$$

and $b(0) = b_0$ say, for some $b : [0,1) \longrightarrow \mathbb{R}^*$. Hence $b(t) = b_0 e^{-\lambda\alpha t}$ and parallel transport here is

$$\tau_t : \overleftarrow{\Pi}_L \circ c(0) \longrightarrow \overleftarrow{\Pi}_L \circ c(t) : (0,b_0) \longmapsto (\alpha t, b_0 e^{-\lambda\alpha t}) .$$

Since $b(t)$ cannot be zero, $c^{\uparrow}(t)$ spirals round one component of LS^1 (if $\alpha \neq 0$), while c proceeds indefinitely round S^1.

Example 5

A connection ∇ in a frame bundle LM determines the geodesic curves, their tangent vectors are parallel along the curves:

$$c : [0,1) \longrightarrow M \text{ with } \nabla_{\dot{c}}\dot{c} = \underline{0} .$$

In local coordinates their differential equations are

$$\frac{d\dot{c}^i}{dt} + \dot{c}^j\dot{c}^k\Gamma^i_{jk} = 0 , \quad i = 1,2,\ldots,n .$$

We return now to a connection ∇ in a principal fibre bundle (P,G,M) to define the holonomy groups of ∇ and here 'curve' will mean a piecewise smooth curve.

The loop space $C(x)$ at any $x \in M$ is the set of all closed curves in M starting and ending at x; there is a natural product on $C(x)$. The holonomy group of ∇ with reference point $x \in M$ is

$$H(x) = \{\tau_c \mid c \in C(x)\}$$

where τ_c is the parallel transport automorphism around c,

$$\tau_c : \overleftarrow{\Pi}_P(x) \longrightarrow \overleftarrow{\Pi}_P(x) .$$

(Note that τ_c commutes with the action of G on the fibre $\overleftarrow{\Pi}_P(x)$.) We can realise $H(x)$ as a subgroup of G, namely

$$H(u) = \{g \in G \mid R_g(u) = \tau_c(u) , \tau_c \in H(x)\} ,$$

the holonomy group of ∇ with reference point $u \in P$.
If we write for $u,v \in P$,

$$u \sim v \iff (\exists \text{ horizontal smooth curve in } P \text{ from } u \text{ to } v)$$

then $\sim$ is an equivalence relation, partitioning P into disjoint non-empty sets and we have

$$H(u) = \{g\in G | u\sim R_g(u)\}$$

Properties of the holonomy group $H(u)$

(i) $(\forall g\in G)\ H(R_g(u)) = ad(g^{-1})H(u)$.

(ii) $u \sim v \Longrightarrow H(u) = H(v)$.

(iii) If M is connected and paracompact then $H(u)$ is a Lie subgroup of G , and $P(u) = \{v\in P | u\sim v\}$ gives a principle fibre bundle over M with structure group $H(u)$. The bundle P is decomposed into the disjoint union of these holonomy bundles $P(u)$, $u\in P$.

(iv) The restricted holonomy group of ∇ with reference point u is the identity component of $H(u)$:

$$H^o(u) = \{g\in H(u) | R_g(u) = \tau_c(u),\ c \text{ homotopic to zero in } C(x)\}.$$

(v) $u \sim v \Longrightarrow H^o(u) = H^o(v)$.

(vi) If M is connected and paracompact then $H^o(u)$ is a connected Lie (normal) subgroup of G and $H(u)/H^o$ is countable.

Proof

Kobayashi and Nomizu [57] p. 71-75, 84-85. □

Remark

The action of G on P preserves horizontality and so $(\forall g\in G)$ we get a bundle isomorphism:

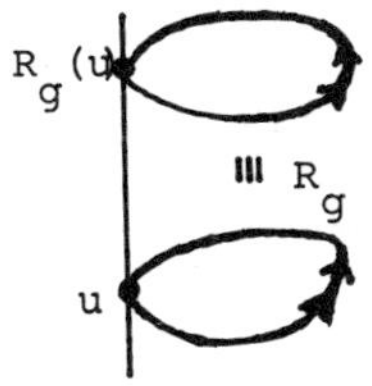

$$R_g : P(u) \longrightarrow P(R_g(u))$$

$$ad(g^{-1}) : H(u) \longrightarrow H(R_g(u)) .$$

Since, $(\forall u,v\in P)(\exists g\in G) : P(v) = P(R_g(u))$,
the holonomy bundles $P(u)$, $u\in P$ are all isomorphic.

Example 6

Consider $L\mathbb{R}$ with the constant connection λ . Horizontal curves are essentially the same as for LS^1 (cf. Example 4). Given any

$$c : [0,1] \longrightarrow \mathbb{R} : t \longmapsto x + \alpha t$$

beginning at x , with α fixed in $\mathbb{R}$, we find

$$c^{\uparrow} : [0,1] \longrightarrow L\mathbb{R} : t \longmapsto (x + \alpha t, be^{\alpha t})$$

for the unique horizontal lift through $(x,b) \in L\mathbb{R}$.
Now, closed curves in $\mathbb{R}$, beginning and ending at $x \in \mathbb{R}$ are essentially of the form

$$c : \begin{cases} t \longmapsto x + \alpha t & t \in [0,\tfrac{1}{2}) \\ t \longmapsto x + \alpha(1-t) & t \in [\tfrac{1}{2},1] \end{cases} .$$

Therefore parallel transport τ_c for $c \in C(x)$ is the identity on $\overleftarrow{\Pi}_L(x)$. The action of $\mathbb{R}^*$ on $L\mathbb{R}$ is free and so, $\forall (x,b) \in L\mathbb{R}$,

$$H(x,b) = \{g \in \mathbb{R}^* \mid R_g(u) = u\} = \{1\} .$$

Hence the holonomy groups are all trivial and the holonomy bundle through (x,b) is simply the exponential curve $y = be^{\alpha t}$, $t \in \mathbb{R}$, (in $L^+\mathbb{R}$ if $b>0$, in $L^-\mathbb{R}$ if $b<0$) . We see that such curves certainly partition $L\mathbb{R}$, since they do not intersect but every point of $L\mathbb{R}$ lies on one of them.

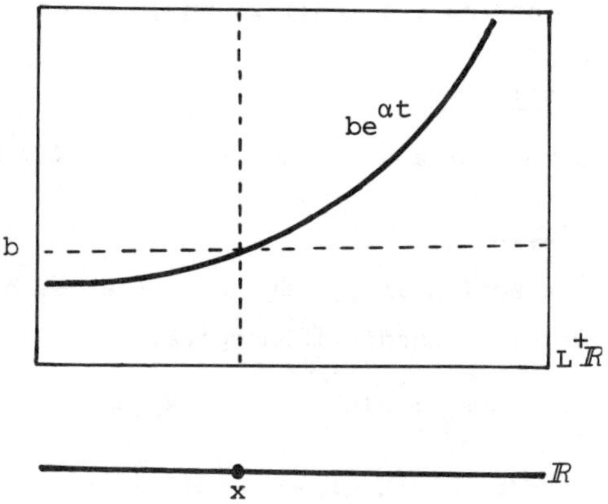

Example 7

Here we see that the holonomy group for LS^1 with constant connection λ is not trivial. For any $x \in S^1 = \mathbb{R} (\text{mod } 1)$ the loop space $C(x)$ still contains trivial curves, as for $\mathbb{R}$, but now we have also non-trivial curves

$$c_k : [0,k] \longrightarrow S^1 : t \longmapsto (x+t) \pmod 1 \quad \text{for any}$$

$k = \pm 1, \pm 2, \ldots$.

It turns out that

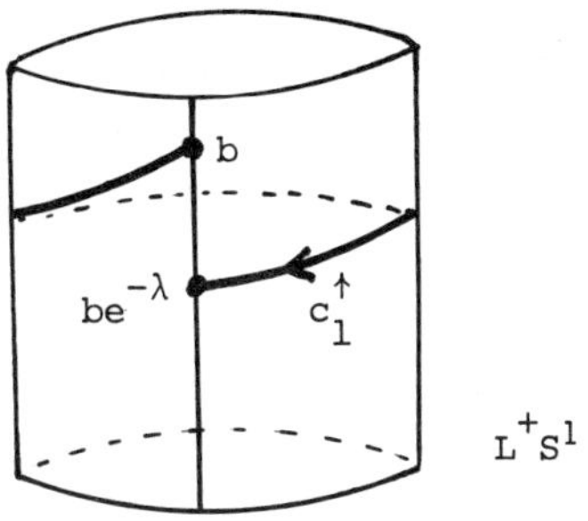

$$H(x) = \{\tau_{c_k} \mid k \in Z\}$$

$$\tau_{c_k} : (x,b) \longmapsto (x, be^{-\lambda k}),$$

so, as a subgroup of $\mathbb{R}^*$ we have

$$H(x,b) = \{e^{-\lambda k} \in \mathbb{R}^* \mid k \in Z\} .$$

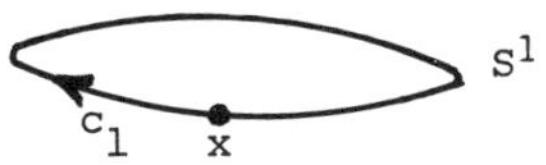

The identity component is trivial, $H^o(x,b) = \{1\}$, so property (vi) is trivially true. The horizontality relation on LS^1 is

$$(x,b) \sim (x, be^{-\lambda k}) \quad \text{if} \quad k \in Z .$$

We can check property (i) above by $(\forall g \in \mathbb{R}^*)$

$$H(R_g(x,b)) = H(x,bg) = H(x,b)$$

$$ad(g^{-1})H(x,b) = H(x,b), \quad \text{(cf. §2.5 Example 2)} .$$

Example 8

The connection in $\mathbb{R}^2$ with:

$$\Gamma^1_{11}(x,y) = y, \ \Gamma^2_{22} = -x, \quad \text{otherwise} \ \Gamma^k_{ij} = 0, \ \text{for all} \ (x,y) \in \mathbb{R}^2$$

is discussed in Dodson [26] p.440. At $(0,0) \in \mathbb{R}^2$ the restricted holonomy group is $\mathbb{R}^+$.

Suppose that (P',G',M) is a reduced bundle (cf. §2.3) of the principle fibre bundle (P,G,M) and that ∇ is a connection in (P,G,M) . Then the connection ∇ is reducible to a connection ∇' in (P',G',M) if the inclusion morphism

$$(P',G',M) \hookrightarrow (P,G,M)$$

preserves the connection structure (horizontal subspaces, connection form, holonomy groups etc., cf. Kobayashi and Nomizu [57] p. 81).

2.6.2 Theorem

Given a connection ∇ in a principal fibre bundle (P,G,M) over a connected paracompact manifold M then each holonomy bundle $P(u)$ is a reduced bundle with structure group $H(u)$ and ∇ is reducible to a connection in $P(u)$.

Proof

Kobayashi and Nomizu [57] p. 84. □

Example 9

We continue Example 7. For any $(x,b)\in L^+S^1$ we have

$$Q = L^+S^1(x,b) = \{(x\hat{+}t, be^{-\lambda t}) \mid t\in\mathbb{R}\}$$

here $\hat{+}$ is addition mod 1 . The projection map is

$$\Pi_Q : (x\hat{+}t, be^{-\lambda t}) \longmapsto x\hat{+}t$$

and the fibre over $x\in S^1$ is

$$\overset{\leftarrow}{\Pi}_Q(x) = \{(x, be^{-\lambda k}) \mid k\in\mathbb{Z}\} .$$

The reduced bundle is then $(Q,\mathbb{Z},S^1)$ where the right action of $\mathbb{Z}$ is

$$Q\times\mathbb{Z} \longrightarrow Q : ((y,d),k) \underset{R_k}{\longmapsto} (y, de^{-\lambda k}) ,$$

evidently a free action since

$$de^{-\lambda k} = d \implies k = 0 .$$

Following §2.3 we need only provide an open cover $\{U_\alpha \mid \alpha\in A\}$ of S^1 and a family $\{\psi_{\alpha\beta} \mid \alpha,\beta\in A\}$ of transition functions to show that Q is a principle fibre bundle to which L^+S^1 is reducible. Take the cover : $U_\alpha = S^1$ $(\forall\alpha\in\mathbb{Z})$

and the diffeomorphisms

$$\psi_\alpha : \overset{\leftarrow}{\Pi}_Q U_\alpha \longrightarrow U_\alpha\times\mathbb{Z} : (r, be^{-\lambda m}) \longmapsto (r, m+\alpha) .$$

Then, putting $(\forall\alpha\in\mathbb{Z})$

$$\phi_\alpha : \overset{\leftarrow}{\Pi}_Q U_\alpha \longrightarrow \mathbb{Z} : (r, be^{-\lambda m}) \longmapsto m+\alpha$$

we have $(\forall k\in\mathbb{Z})$ as required

$$\phi_\alpha \circ R_k = R_k \circ \phi_\alpha .$$

Suppose now that

$$(r, be^{-\lambda m}) \in \overleftarrow{\Pi}_Q(U_\alpha \cap U_\beta) = Q ,$$

then we find

$$\left(\phi_\beta \circ R_k(r, be^{-\lambda m})\right)\left(\phi_\alpha \circ R_k(r, be^{-\lambda m})\right)^{-1}$$

$$= (m+k+\beta) - (m+k+\alpha) = \beta - \alpha$$

$$= \left(\phi_\beta(r, be^{-\lambda m})\right)\left(\phi_\alpha(r, be^{-\lambda m})\right)^{-1} = (m+\beta)-(m+\alpha) = \beta-\alpha .$$

Hence suitable transition functions are (cf. §2.3)

$$\psi_{\beta\alpha} : U_\alpha \cap U_\beta \longrightarrow \mathbb{Z} : r \longmapsto \beta-\alpha ;$$

that is,

$$\psi_{\beta\alpha} \circ \Pi_Q(r, be^{-\lambda m}) = \left(\phi_\beta(r, be^{-\lambda m})\right)\left(\phi_\alpha(r, be^{-\lambda m})\right)^{-1} .$$

This is well-defined because given $r \in \mathrm{dom}(\psi_{\beta\alpha})$ it does not matter to which $(r, be^{-\lambda m})$ we lift it in Q before we apply $\psi_{\beta\alpha} \circ \Pi_Q$.

Now Q is a one-dimensional manifold, an exponential spiral wrapped around the cylinder. Hence at any $(x,b) = u \in Q$ the vertical subspace $\mathbb{Z}_u$ is trivial (cf. §2.3). Hence we have only <u>one</u> choice for a connection ∇' in Q , it has horizontal subspace

$$H_u = T_u Q \quad (\forall u \in Q) .$$

Then as required, since $(\forall k \in \mathbb{Z})$, R_k is linear, $DR_k H_u = H_{R_k(u)}$.

It follows that ∇ (i.e. λ in LS^1) is reducible to ∇' in Q . Evidently Q was constructed from (and only from) the horizontal curves through $(x,b) \in L^+S^1$ so the inclusion $Q \hookrightarrow LS^1$ preserves the horizontal structure. The connection form of ∇' is trivial, because the Lie algebra of $\mathbb{Z}$ is trivial and the holonomy groups of ∇' are again copies of $\mathbb{Z}$.

2.6.3 <u>Theorem</u>

Given a principal fibre bundle (P,G,M) a closed subgroup G' of G and the associated bundle $E = P/G'$ with fibre G/G' , suppose that $\sigma : M \longrightarrow E$ is a section. Then there is a reduced bundle

(P',G',M) corresponding to σ and a connection ∇ in P is reducible to a connection ∇' in P' if and only if σ is parallel with respect to ∇ .

Proof

Kobayashi and Nomizu [57] p.88 . The correspondence : $\sigma \longrightarrow P'$ is one-to-one and derives from the proof of Theorem 2 in §2.3. □

2.6.4 Theorem

Suppose that M is parallelizable by the section

$$p : M \longrightarrow LM : x \longmapsto (p_i)_x$$

then p determines a connection ∇^p in LM such that

$$\nabla^p_{p_i}(p_j) = \underline{0} \quad \forall\ i,j = 1,2,\ldots,n.$$

If

$$h : M \longrightarrow G\ell(n;\mathbb{R}) : x \longmapsto [h^i_j]_x$$

is a smooth map then

$$q : M \longrightarrow LM : x \longmapsto (q_i)_x \ , \ q_i = h^k_i p_k \ ,$$

is another parallelization and

$$\nabla^q = \nabla^p \iff h \text{ is constant on connected components of M.}$$

Proof

The Christoffel symbols, with respect to the frame $(p_k)_x$, for ∇^p all vanish by hypothesis:

$$\nabla^p_{p_i}(p_j) = \Gamma^k_{ij}p_k = \underline{0} \implies \Gamma^k_{ij} = 0 \ \forall\ i,j,k .$$

So from Example 3 we find that locally the splitting of $T_u LM$, $\forall u = \left(x,(b^i_j p_i)_x\right) \in LM$, is given by

$$T_u LM = H_u \oplus G_u$$

$$: \left(x,(b^i_j p_i)_x, X, B\right) = \left(x,(b^i_j p_i)_x, X, 0\right) \oplus \left(x,(b^i_j p_i)_x, 0, B\right) .$$

So we choose

$$H_{p(x)} = D_x p(T_x M)$$

and for any $u \in LM$ with $u = R_h p(x)$

$$H_u = D_{p(x)} R_h (H_{p(x)}) .$$

We know that $LM = M\times G\ell(n;\mathbb{R})$ by the existence of p, now we see that the connection ∇^p makes the horizontal subspaces look horizontal in this product bundle by

$$LM \longrightarrow M\times G : \left(x,(b^i_j p_i)_x\right) \longmapsto \left(x,\delta^i_j\right) ,$$

i.e. we locate the identity in G at the frame determined by the parallelization at each point .

Evidently the given q is another parallelization and

$$\nabla^p_{p_i} q_j = \nabla^p_{p_i} h^k_j p_k = p_i(h^k_j) p_k + h^k_j \nabla^p_{p_i} p_k$$

which equals $\underline{0}$ if and only if h^k_j is constant on each connected component of M . □

Corollary 1

The connection ∇^p need not be symmetric (proved in Theorem 8 §2.8). □

Corollary 2

The geodesics of ∇^p are the integral curves of constant linear combinations like

$$X : M \longrightarrow TM : x \longmapsto a^i p_i . \qquad □$$

2.6.5 Theorem

Suppose that M is a connected, paracompact n-manifold (for some $n>1$) with a principal fibre bundle (P,G,M) then:

(i) There always exists a connection in P, having holonomy bundles coinciding with P .

(ii) If the given bundle is reducible to (P',G',M) then there is a connection in P having each holonomy bundle coinciding with P' .

(iii) If (P',G',M) is a subbundle of (P,G,M) then a connection ∇ in P is reducible to a connection in P' if for all $u\in P'$ the horizontal subspace of T_uP is tangent to P' .

(iv) Any connected Lie group H can be realized as the holonomy group of some connection in $Q = M \times H$.

Proof

(i) A trivial bundle $U \times G$ is constructed over a neighbourhood U of some point in M and provided with a connection having holonomy group the identity component of G . This is extendible to P by means of a partition of unity on M , available because M is paracompact. Details are given in Kobayashi and Nomizu [57] p.90-91 (cf. also Brickell and Clark [9] p. 154), the construction also proves (iv).

(ii) Follows from (i) because the reduced bundle is also a principal fibre bundle.

(iii) Kobayashi and Nomizu [57] p. 85 : define ∇' in P' by $H'_u = H_u$, where H_u is the horizontal subspace determined in T_uP by ∇ and $u \in P' \subseteq P$. □

Corollary

The structure group $G\ell(n;\mathbb{R})$ of LM can be reduced to a subgroup G' if and only if there exists a connection in LM having G' as its holonomy group. (Cf. Theorem 9 in Ch. V §1.)

Proof

By (ii), if LM is reducible then there is such a connection. Conversely, LM is always reducible to the holonomy bundle of a given connection. This result is originally due to Hano and Ozeki [42]. □

From the construction of the holonomy bundle one might expect that it is the 'smallest' reduced subbundle to which a connection is reducible. This is indeed the case as we see in the next result.

2.6.6 Theorem

Suppose that M is a connected, paracompact n-manifold (for some $n>1$) and ∇ is a connection in a principal fibre bundle (P,G,M). If ∇ is reducible to a connection ∇' in a reduced subbundle (P', G', M) then for any $q \in P'$ the holonomy bundle $P(q)$ is a reduced subbundle of P' and ∇' is reducible to ∇^H, the reduction of ∇ to $P(q)$.

<u>Proof</u> (Due to M.J. Slupinski)

We need to show that :

(i) there is a smooth inclusion $h : P(q) \hookrightarrow P'$;

(ii) there is a Lie group map $i : H(q) \hookrightarrow G'$;

(iii) P' is reducible to $P(q)$;

(iv) ∇' is reducible to ∇^H .

(i) Suppose $p \in P(q)$. Then by definition there is a ∇-horizontal curve $\bar{c}$ in P from q to p . This curve projects to give $c = \Pi_p \circ \bar{c}$ in M and c has a unique ∇'-horizontal lift c' in P' . By hypothesis there is a smooth inclusion $j : P' \hookrightarrow P$ which also induces the reduction of ∇ to ∇' . It follows that $j \circ c' = c^{\dagger}$, the unqiue ∇-horizontal lift of c to P and hence $j \circ c' = \bar{c}$. Therefore $p \in P'$ and so $P(q) \subseteq P'$.

(ii) We know that $P(q)$ is a reduced subbundle of P with structure group $H(q)$ and ∇ is reducible to ∇^H , by Theorem 2. We have Lie group inclusions

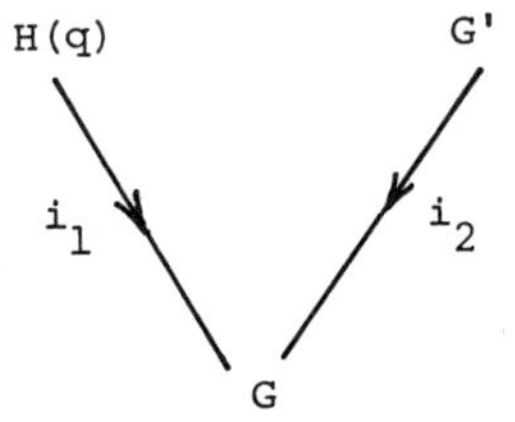

but so far we only have a group inclusion

$$i : H(q) \hookrightarrow G'$$

since $P(q) \subseteq P'$. However, by the Proposition at the end of §2.2 this map i is smooth because the connected components of G' are integral manifolds of the distribution on G defined by the Lie algebra of G', and G' is second countable as a Lie subgroup.

(iii) For this we use Lemma 1 in Kobayashi and Nomizu [57] p. 84.

(iv) We know that ∇ reduces to ∇' in P' and to ∇^H in $P(q)$, and $P(q)$ is a subbundle of P'. Now, if a curve c is ∇'-horizontal in P' through $p \in P(q) \subseteq P'$ then it is ∇-horizontal in P and hence it lies in $P(q)$. So, by Lemma 2 in Kobayashi and Nomizu [57] p. 84, ∇' reduces to some connection in $P(q)$ and by uniqueness of reductions it must be ∇^H. □

2.7 *Differential forms; curvature, torsion*

Certain features of a connection are conveniently expressed in the language of differential forms, that is in terms of purely antisymmetric tensor fields. There is a natural derivation on such fields that preserves anti-symmetry, the exterior derivative. Introductory treatments of the so-called exterior (or Grassmann) algebra and calculus can be found in Spivak [93], Bishop and Goldberg [7], Singer and Thorpe [90], Cartan [11]; further developments and applications are given in Flanders [33] and Kobayashi and Nomizu [57]. Here we shall give about the minimum for our needs, partly because of the excellently detailed treatments mentioned above and partly because it is in the nature of differential forms to use paper extravagantly.

Exterior algebra

Denote by T_r the tensor space of r-linear real valued maps on a vector space V and by ΛT_r denote the vector subspace of purely anti-symmetric members of T_r. Then we have the following

(i) $\dim T_r = n^r$ and $\dim \Lambda T_r = \binom{n}{r}$ if $\dim V = n$.

(ii) There is an alternating operator for each $r = 0,1,\ldots$

$$A_r : T_r \longrightarrow \Lambda T_r : W \longmapsto W_a$$

where for all $v_1, v_2, \ldots, v_r \in V$

$$W_a(v_1,v_2,\ldots,v_r) = \frac{1}{r!} \sum_{\sigma\in S_r} \operatorname{sgn}\sigma\, W(v_{\sigma(1)},v_{\sigma(2)},\ldots,v_{\sigma(r)}) \ .$$

(Here S_r is the symmetric group of permutations on the set $\{1,2,\ldots,r\}$ and $\operatorname{sgn}\sigma$ is the parity homomorphism (cf. Macdonald [65] p. 244) to the multiplicative group on $\{-1,1\}$. Recall that the determinant function is defined similarly - and it is no accident (cf. [28] p. 45 et seq.):

$$\det [a_{ij}] = \sum_{\sigma\in S_r} \operatorname{sgn}\sigma\ a_{1\sigma(1)} \cdot a_{2\sigma(2)} \cdot \ldots \cdot a_{r\sigma(r)} \ .)$$

(iii) There is an associative, anticommutative, distributive <u>exterior product</u>

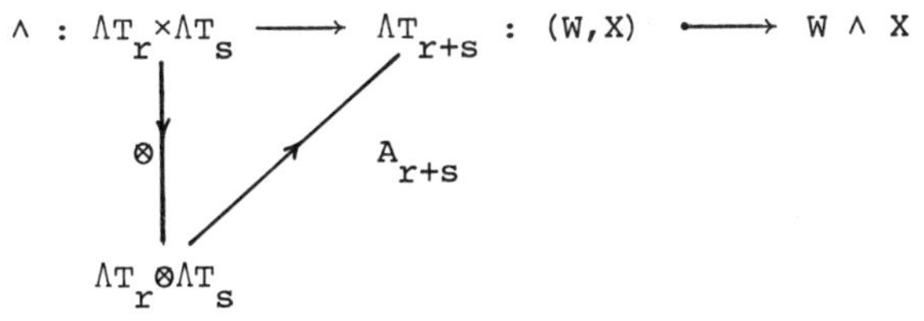

(iv) If (b^i) is a basis for $\Lambda T_1 = T_1$ then a basis for ΛT_r is

$$\{b^{i_1} \wedge b^{i_2} \wedge \ldots \wedge b^{i_r} | i_1 < i_2 \ldots < i_r\} \ .$$

(v) From the properties of binomial coefficients we observe that for each $m = 0,1,2,\ldots,n$

$$\dim \Lambda T_m = \binom{n}{m} = \binom{n}{n-m} = \dim \Lambda T_{n-m} \ .$$

In the presence of a <u>volume element</u> on V (that is any

$$e_1 \wedge e_2 \wedge \ldots \wedge e_n \in \Lambda T_n \ ,$$

where the $(e_i)_{i=1,2,\ldots,n}$ constitute an <u>orthonormal</u> basis) there is an isomorphism for each $m = 0,1,2,\ldots,n$

$$* : \Lambda T_m \longrightarrow \Lambda T_{n-m}$$

called the <u>Hodge star operator</u>. In particular

$$* : \Lambda T_0 \longrightarrow \Lambda T_n : a \longmapsto a\, e_1 \wedge e_2 \wedge \ldots \wedge e_n$$

$$* : \Lambda T_n \longrightarrow \Lambda T_0 : a\, e_1 \wedge e_2 \wedge \ldots \wedge e_n \longmapsto a$$

and quite generally $*$ composes with itself as

$$** = (-1)^{m(n-m)} I_{\Lambda T_m} \qquad (\forall m = 0,1,2,\ldots,n) .$$

Details of the construction can be found in Bishop and Goldberg [7] p. 107 et seq. Note that only in the case n = 3 do we have

$$\binom{1}{3} = \binom{2}{3} = 3$$

and consequently the exterior product on $\mathbb{R}^3$ can be interpreted through the Hodge operator as the vector cross product in $\mathbb{R}^3$, namely (cf. V §2)

$$v_1 \times v_2 = *(v_1 \wedge v_2) .$$

Exterior Analysis

All of the above algebra carries over to the fibres of vector bundles over a manifold. Hence we have the space $\Lambda T_r M$ of antisymmetric $\binom{0}{r}$-tensor fields or (differential) r-forms on M whose elements are sections of the r-form bundle $\Lambda T_r^0 M$. Exterior products for fields are constructed pointwise.

Example 1

Given a chart (U,ϕ) on M then we obtain a frame $(\partial_i)_x$ for $T_x M$, $x \in U$ and its dual $(dx^i)_x$ for $(T_x M)^*$. A frame for e.g. the tensor space $T_2 M_x = (T_x M)^* \otimes (T_x M)^*$ is

$$\{dx^i \otimes dx^j \mid i,j = 1,2,\ldots,n\} \qquad \text{(cf. 1.1)}$$

and a frame for the space $\Lambda T_2 M_x$ is

$$\{dx^{i_1} \wedge dx^{i_2} \mid i_i < i_2;\ i_1,i_2 = 1,2,\ldots,n\}$$

where by definition

$$dx \wedge dy = \tfrac{1}{2}(dx \otimes dy - dy \otimes dx) .$$

The extra ingredient that we have for fields of forms is the (linear) map exterior differentiation

$$d : \Lambda T_r M \longrightarrow \Lambda T_{r+1} M .$$

Amazingly, this very important operator is uniquely characterised by the following properties (cf. Singer and Thorpe [90] p. 108-111 for proof):

(i) d is linear.

(ii) d is the ordinary differential on 0-forms .

(iii) If $\omega \in \Lambda_r TM$, $\sigma \in \Lambda_s TM$ then $d(\omega \wedge \sigma) = d\omega \wedge \sigma + (-1)^r \omega \wedge d\sigma$.

(iv) $d^2 = d \circ d = 0$.

<u>Example 2</u>

Locally $\omega \in \Lambda_r TM$ has an expression

$$\omega = \omega_{(i_1,i_2,\ldots,i_r)} dx^{i_1} \wedge dx^{i_2} \wedge \ldots \wedge dx^{i_r}$$

where the brackets modify the summation convention to act only over increasing sequences $i_1 < i_2 < \ldots < i_r$. Then we find that $d\omega \in \Lambda_{r+1} TM$ has local expression

$$d\omega = d\omega_{(i_1,i_2,\ldots,i_r)} \wedge dx^{i_1} \wedge dx^{i_2} \wedge \ldots \wedge dx^{i_r} .$$

<u>Example 3</u>

In electromagnetic field theory the classical electric and magnetic vector fields can be represented by one differential 2-form F . In regions with negligible matter content Maxwell's field equations are then simply (cf. Clarke [17] and Flanders [33] for more discussion)

$$dF = 0, \quad *d(*F) = J$$

where J is the electric current 1-form .

<u>Example 4</u>

The composites $*d*$ and $*d$ generalise the classical terms 'divergence' and 'curl' from vector calculus. There, many familiar identities are consequences of $d^2 = 0$.

<u>Example 5</u>

If X,Y are vector fields on M and f is a 0-form and ω is a 2-form then $df(X) = X(f)$

$$d\omega(X,Y) = \tfrac{1}{2}\big(X(\omega(Y)) - Y(\omega(X)) - \omega[X,Y]\big) .$$

Similar formulae obtain for higher forms (cf. Kobayashi and Nomizu [57] p.36).

Grassmann homomorphism

Denote by ΛM the Grassmann algebra of all differential forms on M. Then every smooth map $f : M \longrightarrow N$ between manifolds defines an algebra homomorphism $f^* : \Lambda N \longrightarrow \Lambda M$ by putting, for all $\omega \in \Lambda T_r N$, $f^*\omega \in \Lambda T_r M$ such that

$$f^*\omega(v_1, v_2, \ldots, v_r)_x = \omega(Dfv_1, Dfv_2, \ldots, Dfv_r)_{f(x)} ;$$

and if $r = 0$, $f^*\omega = \omega \circ f$.

(What we have here is a contravariant functor at work from Man to the category of Grassmann algebras, cf. II §1.4). It follows that exterior products are preserved and, most significantly, so is exterior differentiation. Precisely we have

(i) $f^*(\omega \wedge \sigma) = f^*\omega \wedge f^*\sigma$

(ii) $f^*d\omega = df^*\omega$.

Proof

Outlined in Singer and Thorpe [90] p. 113 . □

Exactness and closedness

An r-form $\omega \in \Lambda T_r M$ is called:

closed if $d\omega = 0$;

exact if $\exists \sigma \in \Lambda T_{r-1} M : d\sigma = \omega$ (in case $r>0$)

or ω is constant (in case $r=0$) ;

locally exact if $(\forall x \in M)$ ($\exists$ open neighbourhood U of x) : $\omega|_U$ exact.

We observe the implications (since $d^2 = 0$)

exact $\Longrightarrow$ locally exact $\Longrightarrow$ closed .

A deeper consequence is a local converse to

$$\omega = d\sigma \Longrightarrow d\omega = 0 \qquad \text{(true by } d^2 = 0 \text{).}$$

Namely: <u>closed differential forms are locally exact</u>.

The proof is outlined in Bishop and Goldberg [7] p. 175 and depends on an intricate homotopy construction.

<u>Vector valued forms</u>

Hitherto our forms have taken values in $\mathbb{R}$, but any vector space V will serve also. A <u>V-valued r-form</u> ω on M is an assignment to each $x \in M$ of a purely antisymmetric r-linear map

$$\omega_x : T_xM \times T_xM \times \ldots \times T_xM \longrightarrow V .$$

Given any frame $(e_1, e_2, \ldots, e_m)$ for V then ω is expressible as a linear combination of (ordinary) r-forms ω^i on M

$$\omega = \omega^1 e_1 + \omega^2 e_2 + \ldots + \omega^m e_m .$$

Then we define the <u>exterior derivative of the V-valued r-form</u> ω to be

$$d\omega = d\omega^1 e_1 + d\omega^2 e_2 + \ldots + d\omega^m e_m .$$

<u>Example 6</u>

The canonical one-form θ of a frame bundle LM is an $\mathbb{R}^n$-valued 1-form (cf. §2.3).

<u>Example 7</u>

The connection form ω of a connection ∇ in a principal fibre bundle (P,G,M) is a ξ-valued 1-form, where ξ is the Lie algebra of G (cf. §2.5, 2.6). We know that ξ is essentially the matrix space $\mathbb{R}^{n^2}$ if $\dim M = n$ and so locally ω is expressible as an $n \times n$ array of 1-forms (ω^i_j) with values in $\mathbb{R}$. Suppose that $x \in U$ for some chart (U,ϕ) on M and the mutually dual induced frames for T_xM and $(T_xM)^*$ are $(\partial_i)_x$ and $(dx^i)_x$. Then the connection is locally determined by its components Γ^k_{ij} from

$$\nabla_{\partial_i} \partial_j = \Gamma^{\ell}_{ij} \partial_{\ell} \qquad \text{(cf. §2.6)} .$$

Now we find that for each i,j (at $x \in M$)

$$\omega^i_j : T_xM \longrightarrow \mathbb{R} : u \longmapsto dx^i(\nabla_u \partial_j) .$$

But ω^i_j is linear so we can characterise it completely by its action on the frame $(\partial_i)_x$:

$$\omega^i_j(\partial_k) = dx^i(\nabla_{\partial_k}\partial_j) = dx^i(\Gamma^{\ell}_{kj}\partial_{\ell}) = \Gamma^i_{kj} .$$

Thus, each 1-form ω^i_j admits an expression

$$\omega^i_j = \Gamma^i_{kj}dx^k \epsilon (T_xM)^*$$

and so the connection form ω is expressible as the array

$$\omega = [\omega^i_j] = \begin{bmatrix} \Gamma^1_{k1}dx^k & \Gamma^1_{k2}dx^k & \cdots & \Gamma^1_{kn}dx^k \\ \Gamma^2_{k1}dx^k & \Gamma^2_{k2}dx^k & \cdots & \Gamma^2_{kn}dx^k \\ \vdots & & & \vdots \\ \Gamma^n_{k1}dx^k & & \cdots & \Gamma^n_{kn}dx^k \end{bmatrix}$$

Curvature, torsion and the structure equations

Let ∇ be a connection in a principal fibre bundle (P,G,M) with connection form ω . We denote the horizontal projection maps induced by ∇ by

$$\hbar : T_uP = H_u \oplus G_u \longrightarrow H_u : X_H \oplus X_G \longmapsto X_H$$

for all $u \epsilon P$.

The curvature form of ∇ (or of ω) is the 2-form $\Omega = d\omega \circ \hbar$; like ω it takes values in ξ .

The torsion form of ∇ (defined for the case $P = LM$) is the 2-form $\Theta = d\theta \circ \hbar$, where θ is the canonical 1-form, like θ the torsion form takes values in $\mathbb{R}^n$ (cf. §2.3).

The operation on forms σ

$$\sigma \longmapsto d\sigma \circ \hbar$$

is called by Kobayashi and Nomizu [57] p. 77 exterior covariant differentiation and they give details of its properties. The important consequences are as follows, with notation as above:

(i) $\Omega : (X,Y) \longmapsto d\omega(X,Y) + \frac{1}{2}[\omega(X),\omega(Y)]$;

(ii) $d\Omega \circ \bar{h} = 0$;

(iii) $\Theta : (X,Y) \longmapsto d\theta(X,Y) + \frac{1}{2}\big(\omega(X).\theta(Y) - \omega(Y).\theta(X)\big)$;

(iv) $d\Theta \circ \bar{h} = \Omega \wedge \theta$. □

The results (i) and (iii) are called the <u>structure equations</u> (of E. Cartan) and (ii) and (iv) are called the <u>Bianchi identities</u>.

<u>Example 8</u>

We have seen that ω can be viewed as an $(n\times n)$-matrix valued form $[\omega^i_j]$ and similarly Ω is an $(n\times n)$-matrix valued form, $[\Omega^i_j]$ say. Because θ appears as an $(n\times 1)$-matrix valued form, $[\theta^i]$ say, Θ is also an $(n\times 1)$-matrix valued form, $[\Theta^i]$ say .

Then the structure equations are represented by the equations for their typical matrix entries:

(i)' $d\omega^i_j = -\omega^i_k \wedge \omega^k_j + \Omega^i_j$;

(iii)' $d\theta^i = -\omega^i_j \wedge \theta^j + \Theta^i$.

In this notation one readily obtains an identity sometimes called the third structural equation,

$$d\Omega^i_j = d\omega^i_k \wedge \omega^k_j - \omega^i_k \wedge d\omega^k_j ,$$

by applying exterior differentiation to (i)' .

Traditionally in spacetime geometry the notion of curvature was developed via a tensor field, after Riemann. We define this, and another field, and show how they relate to the 2-forms Ω and Θ .

The <u>torsion</u> of a connection ∇ in a frame bundle LM is that $T \in T^1_2 M$ defined by

$$T(X,Y) = \nabla_X Y - \nabla_Y X - [X,Y] , \quad (\forall\ X,Y \in T^1 M)$$

and the <u>curvature</u> of this ∇ in LM is that $R \in T^1_3 M$ defined by

$$R(X,Y)Z = \nabla_X \nabla_Y Z - \nabla_Y \nabla_X Z - \nabla_{[X,Y]} Z ,$$

$$(\forall\ X,Y,Z \in T^1 M) .$$

It follows that any any $x\in M$ with $X,Y,Z\in T_xM$:

$$T(X,Y) = u(2\Theta(\tilde{X},\tilde{Y}))$$

$$R(X,Y)Z = u(2\Omega(\tilde{X},\tilde{Y}))(u^{-1}Z)$$

where $u\in\overleftarrow{\Pi}_L(x)$ is arbitrary and $\tilde{X}$ is any vector in T_uLM which projects by $D\Pi_L$ onto X etc., and we view $u = (u_i)\in\overleftarrow{\Pi}_L(x)$ as the isomorphism

$$u : \mathbb{R}^n \longrightarrow T_xM : (a^i) \longmapsto a^i u_i .$$

Proofs are given in Kobayashi and Nomizu [57] p. 132 et seq., together with further results.

A connection in LM is called <u>symmetric</u> if its torsion is identically zero.

<u>Example 9</u>

Since $T\in T^1_2M$ and $R\in T^1_3M$, they have local expression in terms of the mutually dual frames (∂_i), (dx^i) of the form

$$T = T^i_{jk}\partial_i\otimes dx^j\otimes dx^k$$

$$R = R^i_{jk\ell}\partial_i\otimes dx^j\otimes dx^k\otimes dx^\ell$$

where

$$T(\partial_j,\partial_k) = T^i_{jk}\partial_i$$

and

$$R(\partial_k,\partial_\ell)\partial_j = R^i_{jk\ell}\partial_i .$$

Most books covering elementary differential geometry (e.g. [28]) will show how to deduce the following local expressions for their components in terms of the components of the connection:

$$T^i_{jk} = \Gamma^i_{jk} - \Gamma^i_{kj}$$

$$R^i_{jk\ell} = \partial_k\Gamma^i_{\ell j} - \partial_\ell\Gamma^i_{kj} + \Gamma^h_{kj}\Gamma^i_{\ell h} - \Gamma^h_{\ell j}\Gamma^i_{kh} .$$

We know already that locally

$$\omega^i_j = \Gamma^i_{jk}dx^k$$

and applying d, (cf. Example 2 above)

$$d\omega^i_j = d\Gamma^i_{jk} \wedge dx^k$$

$$= \partial_r \Gamma^i_{jk} dx^r \wedge dx^k .$$

Similarly, we have also

$$\omega^i_h \wedge \omega^h_j = \Gamma^i_{hr} dx^r \wedge \Gamma^h_{jk} dx^k$$

$$= \Gamma^i_{hr} \Gamma^h_{jk} dx^r \wedge dx^k$$

and substitution in (i)' yields

$$\Omega^i_j = \left(\partial_r \Gamma^i_{jk} - \Gamma^i_{hr} \Gamma^h_{jk}\right) dx^r \wedge dx^k ,$$

$$= \tfrac{1}{2} R^i_{jrk}\, dx^r \wedge dx^k$$

by the definition of R^i_{jrk} and the exterior product.

Equivalently, using our modified summation convention for increasing sequences of indices :

$$\Omega^i_j = R^i_{j(rk)}\, dx^r \wedge dx^k .$$

It follows that we can now write the following formulae (cf. e.g. Dodson [24] for details)

$$\omega^i_j \wedge dx^j = T^i_{(kj)} dx^k \wedge dx^j$$

$$d\omega^i_j = -\omega^i_k \wedge dx^k_j + R^i_{j(km)} dx^k \wedge dx^m .$$

It is painfully clear that the actual details of the behaviour of a given connection are rather tedious to calculate. For a manifold of dimension n and a symmetric connection ∇ there are $\tfrac{1}{2}n^2(n+1)$ independent components Γ^k_{ij} and $n^2(n^2-1)/12$ independent components $R^i_{jk\ell}$ of curvature. On the other hand there are only $\binom{n}{2}$ connection forms ω^i_j and the same number of curvature forms Ω^i_j . In a situation where ∇ has a reasonable amount of symmetry it is possible to solve the first Cartan equation for the ω^i_j by inspection. Then the Ω^i_j are readily calculated and yield the components $R^i_{jk\ell}$ by inspection, thus avoiding their direct calculation via the Γ^k_{ij} and their derivatives from the formula given above. The method is due to Cartan and was highlighted by Misner [71]. We shall employ it at the end of the next section.

Remark

The theory of forms proceeds with their integration, leading to the (generalized) theorem of Stokes and to the theorems of De Rham (cf. Bishop and Goldberg [7], Flanders [33] Singer and Thorpe [90], Spivak [93,94] and Hodge [51].)

The theory of connections is really the main theme in Kobayashi and Nomizu [57], which is everyone's main reference. We need one more notion here, that of a flat connection.

A connection ∇ in a principal fibre bundle (P,G,M) is called flat if its curvature form vanishes identically. The canonical flat connection $\underline{\nabla}$ in $P = M\times G$ is defined by the decomposition $\forall u = (x,h) \in M\times G = P$

$$T_u(M\times G) = H_u \oplus G_u$$

with $H_u = T_u(M\times\{h\})$.

So ∇ in $M\times G$ is this $\underline{\nabla}$ if and only if it is reducible to the unique connection in $M\times\{e\}$.

Then it follows that we have

2.7.1 Theorem

(i) A connection ∇ in a principal fibre bundle (P,G,M) is flat if and only if every $x\in M$ has an open neighbourhood U such that there is a bundle isomorphism

$$\psi : \overleftarrow{\Pi}_P U \longrightarrow U\times G$$

mapping the horizontal subspace at each $u\in\overleftarrow{\Pi}_P U$ onto the horizontal subspace at $\psi(u)$ of the canonical flat connection in $U\times G$.

(ii) If M is paracompact and simply connected and ∇ is a flat connection in a principal fibre bundle (P,G,M) then P is isomorphic to the trivial bundle $M\times G$, and the horizontal subspaces of ∇ are mapped onto those of $\underline{\nabla}$ by this isomorphism. (Cf. Theorem 3 in §2.3)

Proof

Kobayashi and Nomizu [57] p. 92-3. □

Corollary 1

A connection ∇ is flat if and only if around every point $x \in M$ there is a neighbourhood N_x such that for any $y \in N_x$ parallel transport gives the same result along any curve in N_x from x to y .

Proof

By (i), if ∇ is flat then each $\overleftarrow{\Pi}_P U$ admits a section p_U that is parallel with respect to ∇ . Locally, the construction resembles that in Theorem 4; namely we choose p_U to satisfy

$$\nabla|_U = \nabla^{p_U} .$$

Then parallel transport is independent of choice of curve in U .

Conversely, if we have some ∇ for which each point x has a neighbourhood N_x in which parallel transport is independent of curve in N_x , then ∇ determines a section p of $\overleftarrow{\Pi}_P N_x$. For, choose any $p(x) \in \overleftarrow{\Pi}_P(x)$ and parallel transport it throughout N_x . Since the map is bijective we obtain a unique $p(y) \in \overleftarrow{\Pi}_P(y)$ for each $y \in N_x$. This allows us to construct the local bundle isomorphisms, required by the theorem to ensure that ∇ is flat, from

$$\psi : \overleftarrow{\Pi}_P U \longrightarrow U \times G : R_h p(y) \longmapsto (y,h) . \qquad \square$$

A direct proof of this corollary for a connection in a frame bundle is given in [28] p. 435 with ample discussion of the geometry involved. Observe that flatness is a local property from either viewpoint: vanishing of curvature differential form or local freedom of parallel transport.

Corollary 2

If ∇ is a flat connection in (P,G,M) over a simply connected manifold M , then parallel transport gives the same result along any curve between each pair of points in M , and P admits a global section.

Proof

If M is simply connected then any two curves from x to y can be continuously deformed one into the other through a family of curves from x to y, i.e. they are homotopic. So we can bridge two such curves by neighbourhoods like N_x, available through Corollary 1 because ∇ is flat.

Parallel transport throughout M yields global sections of P. □

2.8 *Riemannian and pseudo-Riemannian structures*

A metric tensor on a vector space V is a symmetric, non-degenerate bilinear map

$$g : V\times V \longrightarrow \mathbb{R} : (u,v) \longmapsto u\cdot v .$$

(Recall that g is non-degenerate if

$$\big(g(u,v) = 0 \quad (\forall v\in V)\big) \Longrightarrow u = \underline{0} .)$$

It matters particularly whether or not a metric tensor is positive definite i.e. $g(u,u) > 0 \quad \forall u \neq 0$.

An inner product is a positive definite metric tensor.
(Of course a negative definite map can always be converted into a positive definite map by a trivial change of sign, without altering the geometry of the space.)

Example 1

The standard inner product on $\mathbb{R}^n$ is

$$\cdot : \big((x^i),(y^i)\big) \longmapsto x^1y^1 + x^2y^2 + \ldots + x^ny^n .$$

The Lorentz metric tensor on $\mathbb{R}^4$ is

$$\overset{\vee}{\cdot} : \big((x^i),(y^i)\big) \longmapsto -x^1y^1 + x^2y^2 + x^3y^3 + x^4y^4 ,$$

which is not positive definite. (Sometimes the negative of the given map is called the Lorentz metric tensor.)

A metric tensor field on a manifold M is a section of T_2M, i.e. a $\binom{0}{2}$-tensor field $g\in T_2M$, such that $(\forall x\in M)$ g_x is a metric tensor on T_xM .
Such a metric tensor field g is called

(i) A Riemannian structure on M if $(\forall x\in M)$ g_x is an inner product, then (M,g) is a Riemannian manifold ;

(ii) a pseudo-Riemannian structure on M if it is not a Riemannian structure on M, then (M,g) is a pseudo-Riemannian manifold.

Manifolds of type (ii) are particularly relevant to spacetime geometry (cf. V §1, below) but they were sadly neglected by the earlier pure mathematical texts and indeed still often do not have due prominence. The geometry involved in the pseudo-Riemannian case is often much more intricate; for an introductory treatment with examples see Dodson and Poston [28] and Bishop and Goldberg [7].

Example 2

Euclidean n-space is the Riemannian n-manifold $(\mathbb{R}^n, g)$ where, as always using the identity chart for $\mathbb{R}^n$, $(\forall x \in \mathbb{R}^n)$:

$$g_x : T_x\mathbb{R}^n \times T_x\mathbb{R}^n \longrightarrow \mathbb{R} : \big((x,u),(x,v)\big) \longmapsto u \cdot v$$

and $u \cdot v$ denotes the standard inner product on $\mathbb{R}^n$.

Minkowski space is the pseudo-Riemannian 4-manifold $(\mathbb{R}^4, g)$ where $(\forall x \in \mathbb{R}^4)$

$$g_x : T_x\mathbb{R}^4 \times T_x\mathbb{R}^4 \longrightarrow \mathbb{R} : \big((x,u),(x,v)\big) \longmapsto u \overset{\vee}{\cdot} v$$

and $\overset{\vee}{\cdot}$ is the Lorentz metric tensor on $\mathbb{R}^4$.

Example 3

The set of unimodular linear operators on $\mathbb{R}^2$ consists of (2×2)-matrices with determinant 1. In fact it is also a 3-manifold and a Lie group (cf. §1.5), under matrix multiplication; denoted by $S\ell(2;\mathbb{R})$ because it is a subgroup of $G\ell(2;\mathbb{R})$ it has a natural pseudo-Riemannian structure given by

$$g_{\underset{\sim}{x}} : \left(\left(\underset{\sim}{x}, \begin{bmatrix} a & b \\ c & d \end{bmatrix}\right), \left(\underset{\sim}{x}, \begin{bmatrix} p & q \\ r & s \end{bmatrix}\right)\right) \longmapsto \tfrac{1}{2}(as+dp) - \tfrac{1}{2}(br+cq)$$

where we use the identity chart for $S\ell(2;\mathbb{R})$ as a subset of $\mathbb{R}^4$. The geometry of this example is discussed at length in [28] p. 394 et seq.

Example 4

The lecture notes of MacLane [66] on geometrical mechanics develop the theme that, for a physical system, kinetic energy is a Riemannian structure on configuration space.

Example 5

If M is parallelizable by

$$p : M \longrightarrow LM : x \longmapsto (p_i)_x$$

then M admits a Riemannian structure g_p by putting

$$g_p : T_xM \times T_xM \longrightarrow \mathbb{R} : (a^i p_i, b^j p_j) \longmapsto a^i b^j \delta_{ij}$$

where $\delta_{ij} = 1$ if $i = j$ else $\delta_{ij} = 0$. Further consequences are given below in Theorem 6 and its corollaries.

Next we collect some theorems on metric structures.

2.8.1 Theorem

A paracompact manifold M always admits a Riemannian structure.

Proof

A partition of unity allows a smoothing out of a local structure - Brickell and Clark [9] p. 163. The construction fails to adapt to produce indefinite metric tensor fields. In fact the torus and the Klein bottle are the only compact (Hausdorff) 2-manifolds to admit a pseudo-Riemannian structure (cf. Steenrod [96] p. 207). □

Corollary

In particular the result holds for connected M with a countable basis for its topology, since such M have partitions of unity. □

2.8.2 Theorem

A metric tensor field g on a manifold M determines a unique connection ∇^g in LM such that

(i) ∇^g is symmetric (i.e. has torsion zero)

(ii) ∇^g is compatible with g (i.e. parallel transport of tangent vectors is always an isometry; this is equivalent to requiring :

$$u(g(v,w)) = g(\nabla^g_u v, w) + g(v, \nabla^g_u w), \quad \forall u,v,w \in T^1M).$$

Proof

Details are given in e.g. [28] p. 328. Locally, the components Γ_{ij}^{k} of ∇^g are determined by the components $g(\partial_i, \partial_j) = g_{ij}$ of g by :

$$\Gamma_{ij}^{k} = \tfrac{1}{2} g^{km}(\partial_i g_{jm} + \partial_j g_{im} - \partial_m g_{ij})$$

where g^{km} denotes the km-th component of the inverse of $[g_{ij}]$. The connection so determined is called the Levi-Cività or metric connection of g . □

2.8.3 Theorem

Let (M,g) be a connected Riemannian manifold then M is metrizable because we have:

(i) a topological metric or distance function

$$d_g : M \times M \longrightarrow \mathbb{R}$$

$$: (x,y) \longmapsto \inf \left\{ \int \|\dot{c}\| \;\middle|\; c \text{ is a curve from } x \text{ to } y \right\}.$$

(ii) The topology determined by d coincides with the manifold topology on M .

Proof

Details are provided e.g. in Kobayashi and Nomizu [57] p. 157-166. Note that here 'curve' means piecewise C^1-curve and that the integral is the length of the curve i.e.

$$\int \|\dot{c}\| = \int_0^1 \left(g_{c(t)} \left(\dot{c}(t), \dot{c}(t) \right) \right)^{\frac{1}{2}} dt ,$$

where $c : [0,1] \longrightarrow M$ with $c(0) = x$, $c(1) = y$. □

Corollary 1

If (M,d_g) is not complete as a metric space then it can be completed by the standard procedure (cf. e.g., Maddox [68]) of supplying limits for its Cauchy sequences. □

Corollary 2

Like all spaces with a metric topology, M is paracompact. □

We say that a Riemannian manifold (M,g) is complete if the associated metric space (M,d_g) is complete in the sense that every Cauchy sequence is convergent.

2.8.4 Theorem

If (M,g) is a connected Riemannian manifold then

(i) (M,g) is complete if and only if ∇^g is geodesically complete (i.e. every geodesic can be extended to domain $\mathbb{R}$ for an affine parameter, such as arc length). This fails for pseudo-Riemannian g (cf. V §6 below).

(ii) (M,g) is complete if and only if every d_g-bounded subset of M is relatively compact.

(iii) If ∇^g is geodesically complete then any two points x,y in M can be joined by a minimizing geodesic; for x,y close enough this geodesic is unique.

(iv) If all geodesics starting from any particular point of M are infinitely extensible then ∇^g is geodesically complete.

(v) (M,g) is complete if M is compact.

(vi) (M,g) is complete if its group of isometries (g-preserving diffeomorphisms) is transitive on M.

Proof

Kobayashi and Nomizu [57] p. 172 et seq. □

The next result is much newer, apparently discovered independently by Schmidt [84] and Marathe [69], though possibly known to Ehresmann.

2.8.5 Theorem

Let H be the closed subgroup of $G\ell(n;\mathbb{R})$ which leaves invariant a given metric tensor on $\mathbb{R}^n$. Given an n-manifold M denote by E the associated bundle of LM with fibre G/H. Then M is paracompact if E admits a section.

Proof

This is due to Marathe [69] using the following constructions.

(i) Each section $\sigma : M \longrightarrow E$ determines (cf. §2.3 Theorem 2) a unique reduced subbundle P_σ of LM with structure group H .

(ii) A metric tensor on $\mathbb{R}^n$ induces a unique symmetric connection in P_σ and provides M with a pseudo-Riemannian structure g .

(iii) (M,g) determines the Levi-Cività connection ∇^g in LM and hence a Riemannian structure

$$\hat{g} : TLM \times TLM \longrightarrow \mathbb{R}$$

$$: (X,Y) \longmapsto \theta(X)\cdot\theta(Y) + \omega(X)\cdot\omega(Y) \ ,$$

where θ is the canonical 1-form , ω is the connection form of ∇^g and $\cdot$ is the standard inner product on $\mathbb{R}^n$ and $\mathbb{R}^{n^2}$. (We view the Lie algebra ξ as the matrix space $\mathbb{R}^{n^2}$.)

(iv) $(LM,\hat{g})$ is metrizable and hence paracompact. Each component of LM is locally compact and paracompact, so can be written as a countable union of nested compact sets K_n with each K_n contained in the interior of K_{n+1} . Each K_n is metrizable and therefore so is each $\Pi_L(K_n) \subset M$ metrizable. Consequently, M is metrizable and thus paracompact. □

Corollary

M is paracompact if and only if LM admits a connection; in particular, M is paracompact if and only if it admits a pseudo-Riemannian structure.

Proof

This result was also known to Geroch [37] who proved that if M admits a connection (class C^1 will do, since it is existence and openness of the exponential map that is used in his construction) then M has a countable basis for its topology and is therefore paracompact. The converse is obtainable from our Theorems 1 and 2 . □

The metric $\hat{g}$ induced in the frame bundle LM by a connection ∇ is useful in the definition of spacetime singularities, as we shall see in Ch. V §5. In the presence of more structure (a parallelization p) it is possible to achieve an improved resolution of these singularities. The technique is to embed a manifold M in $(LM, \tilde{g}_p)$ where $\tilde{g}_p = \hat{g} + \tilde{p}\cdot\tilde{p}$ i.e.

$$\tilde{g}_p : TLM\times TLM \longrightarrow \mathbb{R}$$

$$: (X,Y) \longmapsto \theta(X)\cdot\theta(Y)+\omega(X)\cdot\omega(Y)+\tilde{p}(X)\cdot\tilde{p}(Y)$$

and $\tilde{p}$ is the connection form induced by the connection ∇^p associated with the parallelization p (cf. Theorem 4 §2.6).

2.8.6 <u>Theorem</u>

Every parallelization p on a connected manifold M determines a unique Riemannian structure g_p on M and a unique Lorentz structure on M.

<u>Proof</u>

The parallelization

$$p : M \longrightarrow LM : x \longmapsto (p_i)_x$$

determines at each $x\in M$ an isomorphism

$$T_xM \equiv \mathbb{R}^n : a^i p_i \longmapsto (a^i) ,$$

which we can make into an isometry by putting

$$g_p(x) : T_xM\times T_xM \longrightarrow \mathbb{R}$$

$$: (a^i p_i, b^j p_j) \longmapsto (a^i)\cdot(b^j) .$$

So we are using the standard inner product $\cdot$ on $\mathbb{R}^n$ to generate via p the Riemannian structure on M:

$$g_p : M \longrightarrow T_2M : x \longmapsto g_p(x) .$$

Since LM consists of <u>ordered</u> bases, we can replace $\cdot$ by the Lorentz structure on $\mathbb{R}^4$. □

<u>Corollary 1</u>

By Theorem 3, g_p determines a unique topological metric d_p on M and this metric topology coincides with the manifold topology, so M is paracompact. □

<u>Corollary 2</u>

By Theorem 2, g_p uniquely determines the Levi-Cività connection ∇^{g_p} and its connection form ω_p. Hence, by Theorem 5, LM has a Riemannian structure

$$\hat{g}_p = \theta\cdot\theta + \omega_p\cdot\omega_p$$

and is therefore paracompact. □

Corollary 3

By construction : for all $x\in M$, if $v\in T_xM$ then $D_xp(v)$ is horizontal with respect to ∇^p and so

$$\tilde{p}(D_xp(v)) = 0, \quad (\forall v\in T_xM) .$$

Hence, $(\forall v,w\in T_xM)$, with $\bar{g}_p = \theta\cdot\theta + \tilde{p}\cdot\tilde{p}$ we have

$$\bar{g}_p(D_xp(v), D_xp(w)) = \theta D_xp(v)\cdot\theta D_xp(w) ,$$

which, in general, differs from $g_p(v,w)$ so the parallelization is not necessarily an isometry. □

Corollary 4

Any parallelization q that is a section of the orthonormal bundle, determined by the metric g_p of a parallelization p , gives a connection ∇^q that (like ∇^p) is compatible with g_p .

Proof

By hypothesis, q gives at each $x\in M$ an orthonormal frame $(q_i)_x$ and so

$$q_k g_p(q_i,q_j) = 0 \qquad (\forall i,j,k) .$$

But also ∇^q satisfies

$$\nabla^q_{q_i} q_j = \underline{0} \qquad (\forall i,j) .$$

So the compatibility condition is satisfied identically. □

We have seen how a parallelization determines a connection, the reverse process also works for simple connections. The idea is to choose a frame at some point and then to determine a frame at other points by parallel transport of the starting frame. Plainly, this will only determine a parallelization if, at least locally, the result of parallel transport does not depend on the choice of curve. So what we need is a flat connection, that is one with zero curvature (discussed at the end of §2.8).

2.8.7 Theorem

(i) Any flat connection ∇ in a frame bundle LM over simply connected M determines a parallelization p that satisfies $\nabla^p = \nabla$; moreover, every parallelization connection is flat.

(ii) If a metric tensor field g on a manifold M determines a non-flat Levi-Cività connection ∇^g and M has a parallelization p that is orthonormal in g , then the connection ∇^p is not symmetric.

Proof

(i) Use the Theorem at the end of §2.8. The flatness of ∇ allows us to bring back to LM the parallelization $M\times\{e\}$ of $M\times G\ell(n;\mathbb{R})$ corresponding to the canonical flat connection. We exploit simple connectedness and use the local bundle isomorphisms. Hence we obtain a parallelization of LM which therefore determines a connection ∇^p . But by construction, the horizontal subspaces defined by ∇^p coincide with those of ∇ . So $\nabla^p = \nabla$.

Any parallelization connection ∇^p is flat by virtue of the same Theorem as in (i). For p stratifies LM globally and hence makes it isomorphic to $M\times G\ell(n;\mathbb{R})$, mapping horizontal subspaces of ∇^p onto those of the canonical flat connection.

(ii) If M has a parallelization p then we know by (i) that ∇^p is flat. Also, LM is trivial and so in the presence of a metric tensor field g we can choose p to lie in a connected component of the orthonormal frame bundle. Hence the compatibility equation is again satisfied by both sides being zero, so ∇^p is compatible with g .

Now, ∇^g is uniquely determined by symmetry and compatibility with g . Hence, if ∇^g is not flat then it must differ from ∇^p which is flat. Therefore ∇^p is not symmetric. □

In Theorem 9 below we shall see that ∇^p is always compatible with g_p , precisely because p is orthonormal in g_p and parallel with respect to ∇^p . We also work out there, in an example, the components of ∇^p and ∇^{g_p} , displaying their compatibilities with g_p .

We saw that a parallelization p uniquely determines a

Riemannian structure g_p on M. This can be useful when (M,g) is a pseudo-Riemannian manifold, for then we have many choices for a Riemannian structure via a partition of unity and a parallelization can help to choose among them. However, if we still wish to make direct use of a given pseudo-Riemannian structure in constructing a Riemannian structure we can do so via a parallelization, as we see next.

2.8.8 Theorem

Let (M,g) be a connected pseudo-Riemannian manifold with a parallelization p. Then there is induced on M a Riemannian structure $\hat{p}$, through the inclusion of pM in the Riemannian manifold $(LM,\hat{g})$.

Proof

Evidently p is a diffeomorphism of M onto pM and so pM is actually a submanifold of LM. Hence $(pM, \hat{g}|_{pM})$ is a Riemannian manifold with $pM \equiv M$ and so a Riemannian structure on M is given by putting

$$\hat{p}(v,w) = \hat{g}(D_x p(v), D_x p(w)) \qquad (\forall v,w \in T_x M). \qquad \square$$

Corollary

Riemannian structures on M are apparently given by isometrically embedding $M \equiv pM$ in :

(i) $(LM,\tilde{g}_p)$ where $\tilde{g}_p = \hat{g}+\tilde{p}\cdot\tilde{p} = \theta\cdot\theta+\omega\cdot\omega+\tilde{p}\cdot\tilde{p}$,

(ii) $(LM,\hat{g}_p)$ where $\hat{g}_p = \theta\cdot\theta+\omega_p\cdot\omega_p$,

(iii) $(LM,\bar{g}_p)$ where $\bar{g}_p = \theta\cdot\theta+\tilde{p}\,\tilde{p}$.

However, on pM we have

$$\tilde{g}_{p|pM} = \hat{g}_{|pM}$$

so the same metric structure is induced on M. Also, $\hat{g}_p$ is a special case of $\hat{g}$, and the parallelization metric g_p on M coincides with that induced by the restriction of $\bar{g}_p$ to pM. $\square$

Of these, only case (i) has dependence on the given pseudo-Riemannian structure, via its Levi-Cività connection form ω.

Also, cases (ii) and (iii) are qualitatively different because ω_p is the Levi-Cività connection form of the parallelization metric structure on M but $\tilde{p}$ is the connection form of the parallelization connection. Our next theorem compares the two connections ∇^p and ∇^{g_p}.

2.8.9 Theorem

A parallelization connection ∇^p is necessarily compatible with the parallelization metric g_p, however it need not be symmetric and therefore need not be the Levi-Cività connection of any Riemannian or pseudo-Riemannian structure. In particular, ∇^p may differ from the Levi-Cività connection ∇^{g_p}.

Proof

We show compatibility directly then the failure of symmetry in a simple example.

(i) The definition of ∇^p for a parallelization

$$p : M \longrightarrow LM : x \longmapsto (p_i)_x$$

is :

$$\nabla^p_{p_i} p_j = \underline{0} \qquad \forall i,j \; .$$

The metric tensor field g_p satisfies (cf. Theorem 7)

$$g_p(p_i,p_j) = 1 \text{ if } i=j, \quad 0 \text{ if } i \neq j \; .$$

Hence, each p_j is a parallel unit vector field and the p_i are mutually orthogonal in g_p. In particular, the requirement for compatibility is

$$p_k(g_p(p_i,p_j)) = g_p(\nabla^p_{p_k} p_i, p_j) + g_p(p_i, \nabla^p_{p_k} p_j) \; .$$

This is its formulation via the Ricci Lemma : that g_p have vanishing covariant derivative (cf. [28] p. 340). It is met identically by both sides vanishing in our case.

(ii) To show that ∇^p need not be symmetric we shall construct a simple counterexample using :

$$p : \mathbb{R}^2 \longrightarrow L\mathbb{R}^2 : (x,y) \longmapsto (e^x \partial_1, e^x \partial_2)$$

where (∂_1, ∂_2) is the standard frame for $T_{(x,y)}\mathbb{R}^2$, via the identity chart on $\mathbb{R}^2$.

This p is a parallelization and it determines a connection ∇^p by the conditions (cf. Theorem 4 §2.6):

$$\nabla^p_{\partial_i} e^x \partial_j = \underline{0} \qquad \text{(for } i,j = 1,2\text{)}$$

where we have used the fact that e^x is never zero, and for any connection ∇, $\nabla_v w$ is linear in v.

Expanding this we find

$$e^x \partial_j + e^x \Gamma^k_{1j} \partial_k = \underline{0} \qquad \text{(for } i = 1\text{)}$$

$$e^x \Gamma^k_{2j} \partial_k = \underline{0} \qquad \text{(for } i = 2\text{)} .$$

Then we can deduce the two matrices of components :

$$[\Gamma^1_{ij}] = \begin{bmatrix} -1 & 0 \\ 0 & 0 \end{bmatrix} \text{ and } [\Gamma^2_{ij}] = \begin{bmatrix} 0 & -1 \\ 0 & 0 \end{bmatrix} .$$

So ∇^p fails to be symmetric because

$$\Gamma^2_{12} \neq \Gamma^2_{21} .$$

Next, p determines a Riemannian structure g_p (via Theorem 6 above) which in standard coordinates has component matrix at $(x,y) \in \mathbb{R}^2$ given by

$$[g_{ij}] = \begin{bmatrix} e^{-2x} & 0 \\ 0 & e^{-2x} \end{bmatrix} .$$

In order to check whether a given connection ∇ is compatible with a given metric tensor field g we can employ Ricci's Lemma (cf. [28] p. 340). It is equivalent to satisfaction of the equation

$$u(g(v,w)) = g(\nabla_u v, w) + g(v, \nabla_u w)$$

for all tangent vector fields u,v,w. We shall use the component form with $u = \partial_k$, $v = \partial_i$, $w = \partial_j$. Taking $g = g_p$, the left hand side is given by the pair of matrices

$$[\partial_1 g_{ij}] = \begin{bmatrix} -2e^{-2x} & 0 \\ 0 & -2e^{-2x} \end{bmatrix} , \quad \partial_2 g_{ij} = \begin{bmatrix} 0 & 0 \\ 0 & 0 \end{bmatrix} .$$

Taking $\nabla = \nabla^p$, the right hand side of the equation is

$$g_p(\Gamma_{ki}^m\partial_m,\partial_j) + g_p(\partial_i,\Gamma_{kj}^m\partial_m)$$

$$= \quad g_{mj}\,\Gamma_{ki}^m + g_{mi}\Gamma_{kj}^m$$

$$= \quad e^{-2x}\Gamma_{ki}^j + e^{-2x}\Gamma_{kj}^i \;.$$

From above :

$$[\Gamma_{1i}^j] = \begin{bmatrix} -1 & 0 \\ 0 & -1 \end{bmatrix}, \quad [\Gamma_{2i}^j] = \begin{bmatrix} 0 & 0 \\ 0 & 0 \end{bmatrix}$$

$$[\Gamma_{1j}^i] = \begin{bmatrix} -1 & 0 \\ 0 & -1 \end{bmatrix}, \quad [\Gamma_{2j}^i] = \begin{bmatrix} 0 & 0 \\ 0 & 0 \end{bmatrix}.$$

By adding these matrices we see that ∇^p is indeed compatible with g_p but since it is not symmetric it is not the Levi-Cività connection of <u>any</u> metric tensor field.

In fact, the Levi-Cività connection ∇^{g_p} determined by the parallelization metric g_p can be found by solving the equation of Ricci's Lemma. In coordinates then, we seek those components $\tilde{\Gamma}_{ij}^k$ satisfying

$$\nabla^{g_p}_{\partial_i}\partial_j = \tilde{\Gamma}_{ij}^k\partial_k \qquad (\text{definition of } \nabla^{g_p})$$

$$g_{km}\tilde{\Gamma}_{ij}^k = \tfrac{1}{2}(\partial_i g_{jm}+\partial_j g_{im}-\partial_m g_{ij}) \qquad (\text{Ricci})\;.$$

It turns out that they are given by

$$[\tilde{\Gamma}_{ij}^1] = \begin{bmatrix} -1 & 0 \\ 0 & 1 \end{bmatrix}, \quad \tilde{\Gamma}_{ij}^2 = \begin{bmatrix} 0 & -1 \\ -1 & 0 \end{bmatrix}.$$

Symmetry is evident and compatibility with g_p is displayed by

$$[\tilde{\Gamma}_{1i}^j] = \begin{bmatrix} -1 & 0 \\ 0 & -1 \end{bmatrix}, \quad [\tilde{\Gamma}_{2i}^j] = \begin{bmatrix} 0 & -1 \\ 1 & 0 \end{bmatrix}$$

$$[\tilde{\Gamma}_{1j}^i] = \begin{bmatrix} -1 & 0 \\ 0 & -1 \end{bmatrix}, \quad [\tilde{\Gamma}_{2j}^i] = \begin{bmatrix} 0 & 1 \\ -1 & 0 \end{bmatrix},$$

for by summing we see that for $k = 1,2$

$$[\partial_k g_{ij}] = e^{-2x}[\tilde{\Gamma}_{ki}^j + \tilde{\Gamma}_{kj}^i]\;. \qquad \square$$

Example

Calculation of Schwarzschild geometry using the Cartan structure equations

This particular example of a pseudo-Riemannian 4-manifold is amply discussed in books on relativity, e.g. Misner, Thorne and Wheeler [72], Clarke [17], Hawking and Ellis [43], Dodson and Poston [28]. Here we set out to find (M,g) where M is topologically $\mathbb{R}^4$ and we suppose it has a global chart yielding coordinates $(x^i) = (r,\theta,\phi,t)$ where $t \in \mathbb{R}$ is time and $(r,\theta,\phi) \in \mathbb{R}^3$ are (pseudo) spherical polar coordinates, valid for $r > 2m$ where m is a positive number yet to be fixed. In fact, (M,g) can successfully represent the geometry of our universe in a region devoid of matter, external to a 'spherically symmetric' mass m and 'static' in time. This can be achieved by seeking that g for which the Ricci tensor vanishes, among the family of Lorentz type metric tensor fields having components

$$g_{11} = f^{-2},\ g_{22} = r^2,\ g_{33} = r^2\sin^2\theta,\ g_{44} = -f^2$$

where f is a real positive function depending on coordinate r only. The normal boundary conditions are $f(r) \longrightarrow 1$ as $r \longrightarrow \infty$, to represent asymptotic flatness of the geometry at large distances from the central mass. For, in the above coordinates the standard Lorentz structure $\bar{g}$ on flat spacetime (Minkowski space) is given by

$$\bar{g}_{11} = 1,\ \bar{g}_{22} = r^2,\ \bar{g}_{33} = r^2\sin^2\theta,\ g_{44} = -1\ .$$

The most efficient use of the Cartan equations is via an orthonormal basis field of 1-forms. Such a basis is, by inspection,

$$(\omega^i) = (f^{-1}dr,\ rd\theta,\ r\sin\theta\ d\phi,\ f\ dt)\ .$$

Now we proceed as follows:

(i) find each $d\omega^i$ and identify each connection form ω^i_j in the first Cartan equation ;

(ii) find the curvature forms by exterior differentiation ;

(iii) extract the $R^i_{jk\ell}$ by inspection ;

(iv) find the components $R_{jk} = R^{i}_{jki}$ of the Ricci tensor.

Step (i)

$$d\omega^1 = d(f^{-1}) \wedge dr = 0 \; ; \quad dr = f\,\omega^1 \,.$$

$$d\omega^2 = dr \wedge d\theta \quad = (f/r)\,\omega^1 \wedge \omega^2 \,.$$

$$d\omega^3 = \sin\theta\, dr \wedge d\phi + \cos\theta\, d\theta \wedge d\phi \,.$$

$$= (f/r)\,\omega^1 \wedge \omega^3 + (\cot\theta)/r\,\omega^2 \wedge \omega^3 \,.$$

$$d\omega^4 = f'dr \wedge dt = f'\,\omega^1 \wedge \omega^4 \; ; \quad f' = \frac{df}{dr} \,.$$

The first Cartan equation, $d\omega^i = -\omega^i_j \wedge \omega^j$, gives:

$$d\omega^4 = -\omega^4_1 \wedge \omega^1 - \omega^4_2 \wedge \omega^2 - \omega^4_3 \wedge \omega^3 \Longrightarrow \omega^4_1 = f'\,\omega^4 \,.$$

$$d\omega^2 = -\omega^2_1 \wedge \omega^1 - \omega^2_3 \wedge \omega^3 - \omega^2_4 \wedge \omega^4 \Longrightarrow \omega^2_1 = (f/r)\omega^2 \,.$$

$$d\omega^3 = -\omega^3_1 \wedge \omega^1 - \omega^3_2 \wedge \omega^2 - \omega^3_4 \wedge \omega^4 \Longrightarrow \omega^3_1 = (f/r)\omega^3 \,,$$

$$\omega^3_2 = (\cot\theta)/r\,\omega^3, \omega^3_4 = 0.$$

Now, since (ω^i) is an orthonormal basis at each point then so is its dual (b_i), hence $g_{ii} = g(b_i, b_i)$ satisfies:

$$g_{11} = g_{22} = g_{33} = 1 \,, \; g_{44} = -1 \,.$$

It follows that since, $\omega^i_j = \Gamma^i_{kj}\,\omega^k$, then

$$\mathrm{sgn}(g_{ii})\,\omega^i_j = -\,\mathrm{sgn}(g_{jj})\,\omega^j_i \,.$$

Hence we deduce that $\omega^3_4 = 0$ implies $\omega^4_2 = 0$ in $d\omega^4$.

The collected results so far are:

$$\omega^2_1 = -\omega^1_2 = (f/r)\omega^2 \quad = f d\theta \,,$$

$$\omega^3_1 = -\omega^1_3 = (f/r)\,\omega^3 \quad = f \sin\theta\, d\phi \,,$$

$$\omega^4_1 = \omega^1_4 = f'\omega^4 \quad = ff'\, dt \,,$$

$$\omega^3_2 = -\omega^2_3 = (\cot\theta)/r\,\omega^3 = \cos\theta\, d\phi \,,$$

$$\omega^4_2 = \omega^2_4 = \omega^4_3 = \omega^3_4 = 0 \,.$$

(Note that we could easily extract the Γ^i_{kj} from these equations.)

Step (ii)

We apply exterior differentiation to the ω^i_j :

$$d\omega^2_1 = df \wedge d\theta = (ff')/r\ \omega^1 \wedge \omega^2 ,$$

$$d\omega^3_1 = (ff')/r\omega^1 \wedge \omega^2 + (f/r)\cot\theta\ \omega^2 \wedge \omega^3 ,$$

$$d\omega^4_1 = \left((f')^2 + ff''\right) \omega^1 \wedge \omega^4 ,$$

$$d\omega^3_2 = (-1/r^2)\ \omega^2 \wedge \omega^3 .$$

The second Cartan equation, $\Omega^i_j = d\omega^i_j + \omega^i_k \wedge \omega^k_j$, gives us the curvature forms by inspection:

$$\Omega^4_1 = d\omega^4_1 + \omega^4_2 \wedge \omega^2_1 + \omega^4_3 \wedge \omega^3_1 = \left(ff' + (f')^2\right) \omega^1 \wedge \omega^4$$

$$\Omega^4_2 = d\omega^4_2 + \omega^4_1 \wedge \omega^1_2 + \omega^4_3 \wedge \omega^3_2 = \left(ff'/r\right) \omega^3 \wedge \omega^4$$

$$\Omega^3_1 = d\omega^3_1 + \omega^3_2 \wedge \omega^2_1 + \omega^3_4 \wedge \omega^4_1$$

$$= \left(ff'/r\right) \omega^1 \wedge \omega^3 + \left(f/r^2\right) \cot\theta\ \omega^2 \wedge \omega^3$$

$$\Omega^2_1 = d\omega^2_1 + \omega^2_3 \wedge \omega^3_1 + \omega^2_4 \wedge \omega^4_1 = \left(ff'/r\right) \omega^1 \wedge \omega^2$$

$$\Omega^3_2 = d\omega^3_2 + \omega^3_1 \wedge \omega^1_2 + \omega^3_4 \wedge \omega^4_2 = \left(f^2-1\right)/r^2\ \omega^2 \wedge \omega^3$$

$$\Omega^4_3 = d\omega^4_3 + \omega^4_1 \wedge \omega^1_3 + \omega^4_2 \wedge \omega^2_3 = \left(ff'/r\right) \omega^3 \wedge \omega^4 .$$

Step (iii)

Next we use $\Omega^k_1 = R^k_{1(ij)}\ \omega^i \wedge \omega^j$, with summation indicated over increasing ij .

Hence, by inspection from Ω^4_1 ,

$$R^4{}_{114} = ff' + (f')^2$$

$$R^4{}_{1ij} = 0 \quad \text{for other } i \leqslant j .$$

Similarly,

$$R^3{}_{113} = (ff'/r) = R^2{}_{112} \text{ and } R^k{}_{1ik} = 0 \quad \text{for } i \neq 1, \forall k;$$

$$R^1{}_{441} = R^4{}_{114} = ff'' + (f')^2 ,$$

$$R^2{}_{442} = R^4{}_{224} = R^4{}_{334} = R^3{}_{443} = (ff'/r) ,$$

$$R^1{}_{331} = R^3{}_{113} = (ff'/r) ,$$

$$R^2{}_{332} = R^3{}_{223} = (f^2/r^2) - (1/r^2) ,$$

$$R^1{}_{221} = R^2{}_{112} = (ff'/r) \ .$$

Step (iv)

From $R_{jk} = R^i_{jki}$ we find that the only surviving terms are the diagonal elements:

$$R_{11} = ff'' + (f')^2 + 2ff'/r$$

$$R_{22} = R_{33} = f^2/r^2 + 2ff'/r - 1/r^2$$

$$R_{44} = ff' + (f')^2 + 2ff'/r \ .$$

Finally, in our particular situation the object was to find the function f such that $[R_{jk}]$ is identically zero. Setting each R_{ii} to zero we find two equations

$$(f^2/r^2) + (2ff'/r) - (1/r^2) = 0 \ ,$$

$$ff' + (f')^2 + (2ff'/r) = 0 \ .$$

However, the first of these implies the second by differentiation, and we can write it in the form

$$f^2 + (f^2)'r - 1 = 0 \quad \text{since} \quad r > 2m > 0 \ .$$

The particular solution that fits observations around the sun is

$$f(r) = \left(1 - (2m/r)\right)^{\frac{1}{2}} \ ,$$

and we see that the curvature components are dominated for small r by $1/r^3$.

V Spacetime structure

The most important reference work on the geometrical physics of spacetime is that of Hawking and Ellis [43]. Our present chapter is intended to complement that work by gathering together various topological results and by presenting the current position on singularities.

By a spacetime we shall mean a connected, non-compact Hausdorff space M with a C^∞ real 4-manifold structure and a C^∞ metric tensor field g of Lorentz type - the latter means that at each $x \in M$ the metric tensor on $T_x M$ is representable as a quadratic form on $\mathbb{R}^4$ with eigenvalues $(-1,1,1,1)$. (See Clarke [13] for the important embedding theorem.)

We shall see that the existence of a Lorentz structure guarantees paracompactness (since M is Hausdorff) and conversely on any non-compact, paracompact Hausdorff manifold there exists a Lorentz structure that does not generate closed timelike curves. It is known that there is a Lorentz type structure on a compact paracompact Hausdorff manifold if and only if the Euler characteristic is zero, but compactness is unphysical in the sense that it leads to closed timelike curves - violating accepted rules of causality. We may as well assume that M is connected since disconnected parts of a universe would presumably not interact. The smoothness of M and g is largely for convenience; indeed, as we pointed out in [28], the bigger step for manifold structures is from C^0 to C^1, rather than from C^1 to C^∞. However, we do need g to determine a connection and hence geodesics, covariant derivatives, exponential maps and parallel transport. The existence theorem at work in each case is essentially the classical Cauchy Theorem for ordinary differential equations (cf. [28] p. 276 and Lang [60]). This assures us of a unique (local) solution if the function to be integrated satisfies a Lipschitz Condition on some open set U ; then, moreover, the solution depends continuously on the initial conditions. In our situation the function to be integrated will depend on the components Γ^k_{ij} of the Levi-Civitȁ connection induced by g (cf. §2.6), and U will be a subset of $\mathbb{R}^4$. A sufficient (but not necessary) requirement for the Lipschitz Condition is that the first partial derivatives of the Γ^k_{ij} exist and are bounded on a convex U . Now, we have seen that the Γ^k_{ij} are determined by the first partial derivatives of the components g_{ij} of the metric tensor field g . So to apply the existence and uniqueness theorem (to geodesics, parallel transport etc.) we need existing and bounded second partial derivatives of the g_{ij} . That is, g must be of class C^{2-} . Now, in order to support a C^{2-} field the differentiable structure on M must be of class C^3 . On the other hand it is known (cf. Munkres [73]) that if $r \geq 1$ then any C^r atlas for M has a C^∞ subatlas, so our resort to the convenience of a C^∞ structure for M is not a serious restriction for most purposes.

We shall suppose that (M,g) is inextensible as a C^∞ Lorentz 4-manifold; that is to say it is not an isometric proper submanifold of any other C^∞ Lorentz 4-manifold (cf. Hawking and Ellis [43] p. 58). The motive for this supposition is that no physically reasonable spacetime should be the result of an arbitrary removal of certain points from a larger spacetime.

We have not made any preparation to study the algebraic topology of spacetime in this book but recent studies in this field have been made, for example by Lee [64] and Whiston [102, 103] . Also, Clarke [14] and Ihrig [53] have used spacetime holonomy groups and much general background material will be found in [22, 23] . We shall not here put jet bundles as such to work, but these will no doubt gain in importance for physical applications (cf. Hennig [44] and Hermann [45]).

The most important Lie group that arises in spacetime theory is the Lorentz group SO(1,3). A thorough study of its representations and physical applications can be found in Gel'fand, Minlos and Shapiro [35], an introduction to its role in relativity is provided in Clarke [17], and Porteous [78] shows how it fits into the general pattern of transformation groups. We collect here a few properties for easy reference.

(i) As a matrix group, we have the representation

$$SO(1,3) = \{\lambda \in G\ell(4;\mathbb{R}) \mid \lambda^T g_o \lambda = g_o\}$$

where λ^T denotes the transpose of λ and g_o is the diagonal matrix with elements $(-1,1,1,1)$; so g_o is actually the quadratic form on $\mathbb{R}^4$ that represents the Lorentz metric tensor (cf. IV §2.8). It follows from the product rule for determinants that

$$\lambda \in SO(1,3) \implies \det\lambda = \pm 1 \ .$$

The proper or orthochronous Lorentz group is the subgroup $SO^+(1,3)$ having elements with determinant +1.

(ii) The notation $SO(1,3)$ emphasises that the Lorentz group consists of Special Orthogonal transformations of Minkowski space i.e. of $\mathbb{R}^4$ with the Lorentz metric tensor g_o ; special means orientation-preserving. In fact, $SO(1,3)$ is an orientation preserving normal Lie subgroup of the orthogonal group $O(1,3)$. Recall that $O(3)$, the group of rotations and reflections of $\mathbb{R}^3$, consists of orthogonal 3×3 real matrices σ satisfying $\sigma^T I \sigma = I$, that is $\sigma^T = \sigma^{-1}$; and $SO(3)$ is the subgroup having determinant +1 , i.e. the rotations.

(iii) $SO(1,3)$ has two components but $SO^+(1,3)$ is connected.

(iv) $SO^+(1,3)$ consists of the rotations of $\mathbb{R}^1 \oplus \mathbb{R}^3$ that preserve the orientations of the two factors.

(v) $SO^+(1,3)$ is non-compact, it has no finite dimensional unitary representations and it has infinite irreducible representations. This is in contrast with the rotation group $SO(3)$ for $\mathbb{R}^3$, which is compact, any representation can be made unitary and any irreducible representation is finite-dimensional. Of course, $SO(3)$ is a subgroup of $SO^+(1,3)$ because for each $\mathbb{R}^3$-rotation $\sigma \in SO(3)$, we obtain a proper Lorentz transformation

$1 \oplus \sigma \in SO^+(1,3)$.

(vi) The <u>universal covering group</u> of $SO^+(1,3)$ is $S\ell(2;\mathbb{C})$ the group of 2×2 complex matrices with determinant $+1$.

(vii) $SO(1,3)$ is a <u>closed</u> subgroup of $G\ell(4;\mathbb{R})$.

(viii) As a manifold, $SO(1,3)$ has <u>dimension 6</u> and hence its Lie algebra is 6-dimensional; the same is true for $SO^+(1,3)$. For comparison: $SO(3)$ has dimension 3, $S\ell(2;\mathbb{C})$ has (real) dimension 6 and $S\ell(2;\mathbb{R})$ has dimension 3.

(ix) The fundamental group of $SO^+(1,3)$ coincides with that of $SO(3)$, it is Z_2 .

1. LORENTZ STRUCTURES

In order to support relativity theory, or any other plausible geometrical theory of physics (cf. Hawking and Ellis [43] and Misner, Thorne and Wheeler [72]) our manifold needs a Lorentz structure if it is to represent the observable universe. The peculiar distinction of a Lorentz structure over other metric structures is that it determines a family (or <u>cone</u>) of directions in tangent spaces, these we call <u>timelike</u> directions. Conversely, if we are given a nowhere-zero vector field

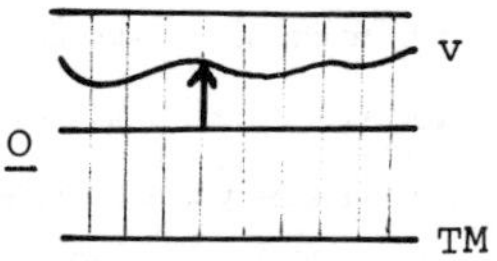

$$v : M \longrightarrow TM$$

on a manifold M , and if M admits a Riemannian structure h then a Lorentz structure h_v is obtained by

$$h_v = h - \frac{2}{h(v,v)} h_\downarrow(v) \otimes h_\downarrow(v)$$

where $h_\downarrow$ is at each $x \in M$ the dualizing isomorphism (cf. [28] for further discussion of $h_\downarrow$ and its inverse $h^\uparrow$)

$$T_xM \longrightarrow (T_xM)^* : a \longmapsto h_x(a,-) = h_\downarrow(a) .$$

This formula for h_v is from Avez [2] and we observe that it is independent of the scale of the field v . So in fact a one-dimensional distribution (cf. IV §2.2), that is a field of unscaled directions (cf. Geroch [39]), or simply line elements (cf.

Steenrod [96] and Markus [70]), would be sufficient to determine h_v from h. Of course there may well be many choices for h, none of which is particularly distinguished so the field v certainly does not determine a unique Lorentz structure. However, with respect to this h_v, the field v is everywhere timelike. Explicitly,

$$h_v(v,v) = h(v,v) - \frac{2}{h(v,v)} h(v,v)h(v,v) = -h(v,v) < 0 ,$$

since any Riemannian h is positive definite. For any Lorentz structure g, a vector $u \in T_x M$ is called (cf. [28]):

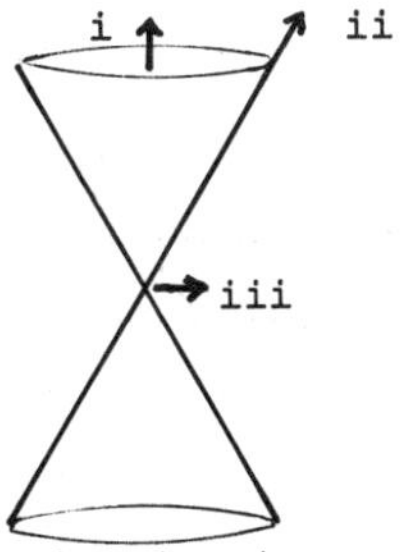

(i) timelike if $g_x(u,u) < 0$;

(ii) null or lightlike if $g_x(u,u) = 0$;

(iii) spacelike if $g_x(u,u) > 0$.

Now, conversely, we can work from a Lorentz structure g on M to find a field of timelike directions, provided that M also has a Riemannian structure h. For example, from Avez [2] (cf. Geroch [39] p. 79 for a picture of what is happening) consider

$$v_h : M \longrightarrow TM : x \longmapsto \pm v_h(x)$$

where $v_h(x) \in T_x M$ is that unique (up to sign) element u with $h_x(u,u) = 1$ which maximizes $-g_x(u,u)$.

In summary then we have:

1.1 Theorem

Let (M,h) be a Riemannian manifold, then

(i) every one-dimensional distribution on M determines a Lorentz structure g_v on M with respect to which v is timelike;

(ii) every Lorentz structure g on M determines a timelike distribution on M. □

Corollary 1

By Theorem 1 in §2.8, the properties (i) and (ii) will also hold for any paracompact manifold M because these always admit Riemannian structures. □

Corollary 2

By the Corollary to Theorem 5 in §2.8, the properties (i) and (ii) again hold for any manifold M having a connection in its frame bundle, because that is a (necessary and) sufficient condition for paracompactness. □

Corollary 3

Since a manifold M is paracompact if and only if each of its components is second countable (has a countable base for its topology) properties (i) and (ii) hold for second countable manifolds. A sufficient condition for second countability is the admission of a countable atlas. (cf. Brickell and Clark [9] p. 43 et seq.) □

Corollary 4

Since a parallelization of an n-manifold is a specification of n linearly independent vector fields, a parallelizable manifold has n nowhere-zero vector fields and hence many Lorentz structures. In Theorem 6 of Ch. IV §2.8 we saw how one of these could be selected in a unique way. □

We conclude that a nowhere-zero vector field (or at least a field of directions) and paracompactness are needed to determine a Lorentz structure. Conversely, a manifold is paracompact if and only if it admits a Lorentz structure, by the Corollary to Theorem 5 in §2.8, so we have

1.2 Theorem

Every spacetime is paracompact. □
(A direct proof is given in Geroch [37] p. 1743-4.)

Corollary 1

Every spacetime is second countable, first countable and separable (cf. III §1.5). □

Corollary 2

Every spacetime admits a Riemannian structure and therefore it has a metric topology, by Theorems 1 and 3 in §2.8. □

Corollary 3

Every spacetime admits a partition of unity.

Proof

Brickell and Clark [9] p. 51. □

Corollary 4

If U is an open subset of a spacetime M and V is a closed set contained in U then there exists a smooth real function f on M taking the value 1 on V and the value 0 on $M\backslash U$.

Proof

Brickell and Clark [9] p. 51, via a partition of unity. A similar device is commonly used to smooth out functions given locally on a coordinate neighbourhood. □

We shall require our spacetimes to be non-compact for good physical reasons. However, spacetime manifolds are often constructed as products involving compact manifolds so we note the following.

1.3 Theorem

Any compact manifold admits a nowhere-zero vector field if and only if its Euler characteristic is zero.

Proof

Husemoller [52] p. 259, (cf. also Markus [70]). □

Corollary 1

The Euler characteristic of the sphere S^n is $1+(-1)^n$ so only odd dimensional spheres admit nowhere-zero vector fields. Hence, for this reason alone, S^4 cannot be provided with a Lorentz structure, but $S^3 \times S^1$ could be so provided. (Both are of course inextensible.) □

Corollary 2

Any compact odd dimensional manifold admits a nowhere-zero vector field because its Euler characteristic is zero.

Proof

Steenrod [96] p. 207. □

1.4 Theorem

If any manifold M_1 admits a nowhere-zero vector field then so does $M_1 \times M_2$ for any manifold M_2.

Proof

Brickell and Clark [9] p. 118. □

Corollary 1

For any manifold M_o, $\mathbb{R} \times M_o$ always admits a nowhere-zero vector field because $\mathbb{R}$ does. □

Corollary 2

The product manifold $\mathbb{R} \times S^3$ can be given the structure of a spacetime (it and $\mathbb{R} \times \mathbb{R}^3$ are the most popular candiates for cosmological models, taking $\mathbb{R}$ as the time coordinate). □

The problem with compact manifolds is seen in the next result.

1.5 Theorem

If g is a Lorentz structure on a compact 4-manifold M then there exist closed curves in M having timelike tangent vectors everywhere.

Proof

Geroch [36], (cf. also Geroch [39] p. 81 for a picture, and Bass and Witten [3]); Tipler [98] gives a stronger result. □

Corollary

The manifold $S^1 \times S^3$ certainly admits a Lorentz structure, but is unacceptable physically because it is compact and will contain a closed timelike curve; the same is true of $S^1 \times S^1 \times S^1 \times S^1$ for example. □

From Sard's Theorem (in IV §1.3) we know that the set of singular points of a smooth map $f : M \longrightarrow \mathbb{R}$ has measure zero. But $x \in M$ is such a singular point if at x the 1-form field df takes the value $\underline{0} \in T_x M^*$. We get an easy deduction:

1.6 Theorem

Every smooth map $f : M \longrightarrow \mathbb{R}$ determines on a paracompact manifold M a vector field that is nowhere-zero, except on a set of measure zero.

Proof

By hypothesis, df is a section of the cotangent bundle and by Sard's Theorem it is zero on a set of measure zero. If this set contains a point x then in some chart (U,ϕ) about x we have:

$$df = \partial_i f \, dx^i \quad \text{on } U$$

and

$$(\partial_i f) = \underline{0} \in \mathbb{R}^n \quad \text{at } x .$$

Hence we obtain a section $df^\dagger$ of the tangent bundle by attaching the components $(\partial_i f)$ of df to each frame (∂_i) dual to (dx^i). That is, about $x \in U$,

$$df^\dagger_U = f^i \partial_i \qquad \begin{cases} \text{where } f^i = \partial_i f \ \forall i \\ \text{and } dx^j(\partial_i) = \delta^j_i \ \forall i,j . \end{cases}$$

and we use paracompactness to smooth out these local sections via a partition of unity. □

Now we concentrate on non-compact manifolds. There we can always provide a Lorentz structure in the presence of paracompactness, because of the following results.

1.7 Theorem

Every non-compact paracompact manifold admits a nowhere-zero vector field.

Proof

Markus [70] Steenrod [96]. (Cf. Hirsch [49] and also Geroch [39] p. 80 for an intuitive account of how to push isolated singularities 'out to infinity', i.e. outside any compact set.) □

Corollary 1

Every non-compact paracompact manifold M admits a Lorentz structure by Theorem 1. Then the frame bundle LM is reducible to OM, the principal fibre bundle of orthonormal frames, with structure group the Lorentz group SO(1,3).

Proof

Markus [70] p. 414. Cf. Chapter IV, Theorem 2.3.2 and note that the Lorentz group is closed in $G\ell(4;\mathbb{R})$. □

Corollary 2

On a non-compact paracompact 4-manifold a Lorentz structure can always be so chosen as to give no closed timelike curves.

Proof

Penrose [75]. □

We can bring Corollary 1 to bear on some results in the previous chapter to express the next deductions.

1.8 Theorem

The frame bundle LM over a spacetime (M,g) is reducible to a principal fibre bundle, with structure group G' a connected Lie subgroup of the Lorentz group, if and only if the associated bundle with fibre $G\ell(4;\mathbb{R})/G'$ admits a section

$$\sigma : M \longrightarrow LM/G'$$

Proof

This is Theorem 2 in Ch. IV §2.3, with the added ingredient:

Lemma The connected Lie subgroups of the Lorentz group are closed in $G\ell(4;\mathbb{R})$.

Proof This result is given in Schmidt [83] and discussed by Friedrich [34]. The following approach was suggested to me by D.B. Epstein: From Hochschild [50] p. 192, if G' is a connected Lie subgroup of G then G' is closed in G $\Longleftrightarrow$ the closure of every 1-parameter subgroup is contained in G'. A 1-parameter subgroup is either closed or its closure is a torus (cf. [50] p.191).

Now, the Lorentz group is itself closed in $G\ell(4;\mathbb{R})$ so it is sufficient to show that connected Lie subgroups of the Lorentz group are closed. Any torus is contained in a maximal compact subgroup, this in the case of the Lorentz group is the special orthogonal group SO(3) which has maximal torus a circle (cf. Husemoller [52] p. 182). Hence, every 1-parameter subgroup of the Lorentz group is either a circle or a closed copy of $\mathbb{R}$, and it is therefore closed in the Lorentz group and in $G\ell(4;\mathbb{R})$. □

1.9 Theorem

The structure group $G\ell(4;\mathbb{R})$ of LM for a spacetime (M,g) can be reduced to a subgroup G' if and only if there exists a connection in LM having G' as its holonomy group. Therefore, LM admits a connection having the Lorentz group as its holonomy group.

Proof

This is a consequence of Theorem 5 in Ch. IV §2.6, because of Corollary 1 to Theorem 7 in our present section. □

2. ORIENTABILITY

We recall that a frame (x,b) at a point x in a manifold M is an ordered basis $b_1, b_2, \ldots, b_n$ for T_xM. The set of such frames at x is the fibre $\overleftarrow{\Pi}_L(x)$ of the frame bundle LM (cf. IV §2.3). Any two such frames (x,b), (x,b') are related through a real non-singular matrix $[h^i_j]$ by

$$b'_j = h^i_j b_i \ .$$

Fundamental to the notion of orientability is the equivalence relation $\sim x$ on $\overleftarrow{\Pi}_L(x)$ given by :

$$(x,b') \sim x \ (x,b) \iff b'_j = h^i_j b_i \quad \text{with} \quad \det[h^i_j] > 0 \ .$$

It partitions $\overleftarrow{\Pi}_L(x)$ into two equivalence classes and each class is called an orientation of T_xM. A chart (U,ϕ) about any $x \in M$ determines a frame $(\partial_i)_x = (\partial_1, \partial_2, \ldots, \partial_n)_x$. If also x lies in the domain of another chart (U',ϕ') giving a frame $(\partial'_i)_x$ then the non-singular matrix function relating them on $U \cap U'$ is the Jacobian matrix; if it has positive determinant everywhere on $U \cap U'$

then we say that the two orientations, $[(\partial_i)_x]_{\sim x}$ and $[(\partial_i')_x]_{\sim x}$, agree for $x \in U \cap U'$. In that case we obtain an orientation of $T_y M$ for all $y \in U \cap U'$; that is, an orientation on $U \cap U'$.

A manifold M is orientable if it admits an atlas giving an orientation on the whole of M; with such a choice of atlas M is called an oriented manifold. Like dimension, orientability is invariant under diffeomorphisms.

Evidently, if M admits a (continuous) section of LM then it is orientable, just as if it had an atlas consisting of one global chart (as does $G\ell(n;\mathbb{R})$ for example) so we have:

2.1 Theorem

Every parallelizable manifold is orientable. □

Corollary

The following are parallelizable and so orientable:

(i) every open submanifold of $\mathbb{R}^n$;

(ii) every Lie group;

(iii) the spheres S^1, S^3 and S^7;
(cf. below)

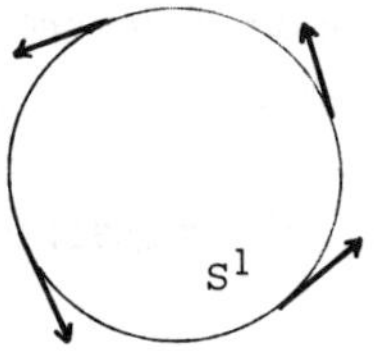

(iv) a product of two or more spheres if at least one has odd dimension (cf. Hirsch [49] p. 98, and Examples below).

(v) every product of parallelizable manifolds. □

The situation is also straightforward if M admits an atlas consisting of just two charts:

2.2 Theorem

If $\{(U,\phi), (U',\phi')\}$ is an atlas for M and $U \cap U'$ is connected then M is orientable.

Proof

The Jacobian determinant function cannot change sign on $U \cap U'$. We can therefore find a two chart atlas making this determinant positive, if necessary by changing the sign of one of the coordinates determined by (U',ϕ'). □

Corollary

All spheres are orientable because they admit the two-chart atlas of stereographic projections (cf. Porteous [78] p. 169). □

There is an equivalent specification of orientability as follows (cf. Singer and Thorpe [90] p. 128). Suppose that M has an atlas $\{(U_\alpha,\phi_\alpha)\,|\,\alpha\in A\}$. Then M is orientable if it admits a subatlas indexed by A' such that whenever (x^i), (y^i) are coordinates with respect to ϕ_α, ϕ_β on $U_\alpha\cap U_\beta$ for $\alpha,\beta\in A'$ then the real function f on $U_\alpha\cap U_\beta$ defined by

$$(dx^1 \wedge \dots \wedge dx^n)_x = f(x)(dy^1 \wedge \dots \wedge dy^n)_x$$

is everywhere positive. We observe that this is an equation between n-forms on $U_\alpha\cap U_\beta$ (cf. IV §2.7). In fact, orientability is completely characterized by the existence of a nowhere-zero n-form on M, also called a volume element on M (cf. IV §2.7):

2.3 Theorem

Let M be an n-manifold; then

(i) M is orientable if it admits a volume element,

(ii) M has a volume element if it is orientable and paracompact.

Proof

Singer and Thorpe [90] p. 129-134. □

The great advantage of this result is that it allows non-orientability to be revealed by showing that every (continuous) n-form is zero somewhere, usually a much simpler process than showing the non-existence of an oriented atlas. There is a standard volume element determined by the metric tensor field on any Riemannian or pseudo-Riemannian oriented manifold, and its relation to the Hodge star operator (cf. IV §2.7) is indicated next.

2.4 Theorem

Given an oriented n-manifold M and on it a metric tensor field g with eigenvalues

$$\underbrace{(-1, \ldots, -1,}_{b} \; \underbrace{1, \ldots, 1)}_{n-b}$$

then the standard volume element is given in local coordinates by

$$v = |\det g_{ij}|^{\frac{1}{2}} dx^1 \wedge \ldots \wedge dx^n$$

where $\qquad g = g_{ij}\, dx^i \otimes dx^j$.

The following results hold:

(i) $*1 = v$, (here of course 1 is a constant element of $\Lambda T_o M = \mathbb{R}$) ;

(ii) $**1 = *v = (-1)^b$;

(iii) for any r-form w, $**w = (-1)^{r(n-r)+b}\, w$;

(iv) g determines a unique dual $g^\dagger \in T^2M$ and for all 1-forms λ, μ it follows that

$$*(\lambda \wedge *\mu) = g^\dagger(\lambda,\mu)**1 = (-1)^b\, g^\dagger(\lambda,\mu) .$$

<u>Proof</u>

(i) By non-degeneracy of g it follows that in any coordinates $\det g_{ij} \neq 0$, hence v is a volume form. Then $*1 = v$ is given by definition.

(ii) Since $*$ is linear, and we can work locally

$$\begin{aligned} **1 = *v &= |\det g_{ij}|^{\frac{1}{2}} *(dx^1 \wedge \ldots \wedge dx^n) \\ &= |\det g_{ij}|^{\frac{1}{2}} |\det g_{ij}|^{\frac{1}{2}} (\det(g_{ij})^{-1}) \qquad \text{(cf. (iii))} \\ &= |\det g_{ij}| (\det(g_{ij})^{-1}) = (-1)^b . \end{aligned}$$

(iii) The Hodge operator has the following action on any r-form $w = w_{(i_1 \ldots i_r)} dx^{i_1} \wedge \ldots \wedge dx^{i_r}$ (cf. Flanders [33])

$$*w = w_{(j_1 \ldots j_{n-r})} dx^{j_1} \wedge \ldots \wedge dx^{j_{n-r}}$$

where

$$w_{j_1 \ldots j_{n-r}} = |\det g_{ij}|^{\frac{1}{2}}\, w^{(k_1 \ldots k_r)} \; \mathrm{sgn}(k_1, \ldots, k_r, j_1, \ldots, j_{n-r})$$

and

$$w^{k_1 \ldots k_r} = g^{k_1 i_1} \ldots g^{k_r i_r} w_{i_1 \ldots i_r}$$

with $(g_{ij})^{-1} = (g^{ij})$.

Now, by the product rule for determinants

$$(\det g_{ij})(\det g^{ij}) = 1 ,$$

also

$$\det g^{ij} = \sum_{(i_1 \ldots i_n)} \operatorname{sgn}(i_1, \ldots, i_n)\, g^{1 i_1} \ldots g^{n i_n} .$$

The result follows, after some manipulation.

(iv) The construction of $g^{\dagger}$ is discussed in Dodson and Poston [28]. Its components are $(g^{ij}) = (g_{ij})^{-1}$. Given $\lambda = \lambda_i dx^i$, $\mu = \mu_j dx^j$ we have

$$g^{\dagger}(\lambda, \mu) = g^{ij} \lambda_i \mu_j .$$

From (iii)

$$*\mu = |\det g_{ij}|^{\frac{1}{2}} \sum_m g^{mj} \mu_j \operatorname{sgn}(m, i_1, \ldots, i_{n-1}) dx^{i_1} \wedge \ldots \wedge dx^{i_{n-1}}$$

and so

$$\lambda \wedge * \mu = |\det g_{ij}|^{\frac{1}{2}} g^{ij} \lambda_i \mu_j \, dx^1 \wedge \ldots \wedge dx^n$$

hence the result. □

<u>Corollary 1</u>

In the case $n = 3$ and $M = \mathbb{R}^3$ with the standard metric structure $(b = 0)$, the cross product $\times$ for geometrical vectors $\lambda, \mu \in \Lambda\mathbb{R}^3 = \mathbb{R}^3$ is related to $\wedge$ by :

$$*(\lambda \wedge \mu) = \lambda \times \mu . \qquad \square$$

<u>Corollary 2</u>

In the familiar case of M being a 2-dimensional oriented submanifold of $\mathbb{R}^3$ we can see the role of the standard volume element on (M,g), where g is the Riemannian structure induced on M by the Euclidean geometry of $\mathbb{R}^3$. Given a chart (U,ϕ) on M with corodinates (x^1,x^2), the usual area of the set U

is

$$\text{Area }(U) = \iint_{(x^i)\in\phi U} |\det g_{ij}|^{\frac{1}{2}} \, dx^1 dx^2 \ .$$

This expression generalises to subsets of $\mathbb{R}^n$ referred to arbitrary coordinates. Since the components of g change under a local coordinate change $(x^i) \longmapsto (\bar{x}^i)$ according to

$$g_{ij} dx^i \otimes dx^j = \bar{g}_{ij} d\bar{x}^i \otimes d\bar{x}^j$$

we can also see why the Jacobian determinant arises in the elementary treatments of multiple integrals. □

Remark

Applications of the volume form in spacetime geometry are given in [72] and [101].

Examples

(i) The following are orientable manifolds:

(a) TM for any manifold M (cf. Greub, Halperin and Vanstone [40] p. 129) ;

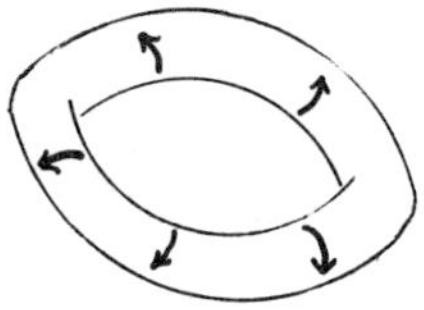

(b) the torus $S^1 \times S^1$;

(c) every 1-manifold ;

(d) the product of orientable manifolds ;

(e) real projective n-space $\mathbb{R}P^n$ for all odd n .

(ii) The following are non-orientable manifolds:

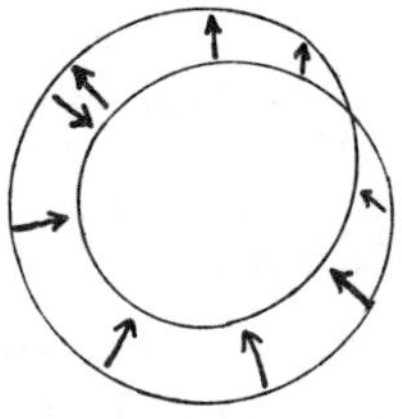

(a) the Mobius strip ;

(b) the Klein bottle ;

(c) the product of any manifold with a non-orientable manifold ;

(d) real projective space $\mathbb{R}P^n$ for any even $n \geq 2$.

For more discussion see Singer and Thorpe [90] and Spivak [94] .

The next result shows how to construct an orientable covering manifold over a non-orientable manifold.

2.5 Theorem

If M is a connected non-orientable manifold then there exists a connected orientable manifold $\tilde{M}$ which is a twofold covering of M.

Proof

(After Schwartz [87] p. 14.)

Let $\{(U_\alpha, \phi_\alpha)\,|\,\alpha \in A\}$ be an atlas for M and define

$$S = \{(x, U_\alpha)\,|\,\alpha \in A,\ x \in U_\alpha\}\ .$$

We take two copies of S, S_+ and S_- say. Denote by $|J_{\alpha\beta}|_x$ the determinant of the Jacobian matrix of the change of coordinates from chart α to chart β at $x \in U_\alpha \cap U_\beta$. We introduce a relation $\sim$ on $S_+ \cup S_-$ by :

$$(x,U_\alpha)_+ \sim (y,U_\beta)_+ \iff x=y \text{ and } |J_{\alpha\beta}|_x > 0\ ;$$

$$(x,U_\alpha)_+ \sim (y,U_\beta)_- \iff x=y \text{ and } |J_{\alpha\beta}|_x < 0\ ;$$

$$(x,U_\alpha)_- \sim (y,U_\beta)_- \iff x=y \text{ and } |J_{\alpha\beta}|_x > 0\ .$$

This is an equivalence relation and its equivalence classes constitute the elements of $\tilde{M}$. A topology is provided by taking, for example,

$$\{[(y,U_\alpha)_+]_\sim \,|\, y \in U_\alpha\}$$

as a neighbourhood of $[(x,U_\alpha)_+]_\sim \in \tilde{M}$.

A manifold structure is provided by :

(i) about $[(x,U_\alpha)_+]_\sim$ use the coordinates $(x^1,x^2,\ldots,x^n)$ of ϕ_α

(ii) about $[(x,U_\alpha)_-]_\sim$ use the coordinates $(x^1,x^2,\ldots,x^{n-1},-x^n)$.

It follows that $\tilde{M}$ is orientable, and therefore connected for essentially the same argument as in Theorem 4. Also,

$$f : \tilde{M} \longrightarrow M : [(x,U_\alpha)_\pm]_\sim \longmapsto x$$

is a two to one, differentiable, covering map. □

We have seen that on a paracompact manifold M the existence of a Lorentz structure is equivalent to the existence of a 1-dimensional distribution on M. Such a distribution is called an orientable line element field if it is generated by a continuous tangent vector field; in that case there is available a continuous assignation of a 'forward' direction at each point. A necessary and sufficient condition for a line element field to be orientable is given by the following result of Markus [70].

2.6 Theorem

A continuous line element field L on a paracompact, separable manifold M is orientable if and only if the orientation of L is preserved around each of a set of generators of $\pi_1(M)$, the first homotopy group (cf. III §1.7). If L is not orientable then M has a twofold covering $\tilde{M}$ and L can be lifted to an orientable line element field $\tilde{L}$ on $\tilde{M}$. If $\pi_1(M)$ contains no proper subgroup of index two, then every continuous line element field on M is orientable. □

(The index of a subgroup H of a group G is the number of (right or left) cosets of H in G.)

Markus [70] also pointed out that if $\pi_1(M)$ contains no proper subgroups of index two then every line element field L that is continuous except for isolated singularities is orientable if $n \geq 3$. For in this case $\pi_1(M)$ is isomorphic to the fundamental group of the open submanifold of M on which L is defined and continuous. Therefore L is always orientable on a simply-connected manifold (which is then its own universal covering manifold), or if $n \geq 3$, in the locality of an isolated singularity.

We shall say that a spacetime (M,g) is time-orientable (also called isochronous) if the line element field determined by g (Theorem 1 in §1) is orientable. Following from the results of Markus we deduce:

2.7 Theorem

Every spacetime $(\tilde{M},\tilde{g})$ admits a covering spacetime (M,g) which is time-orientable. □

Remarks

(Cf. Geroch [36], where $(\tilde{M},\tilde{g})$ is called the Lorentz covering manifold.)

(i) We actually construct $\tilde{M}$ from an equivalence relation on the set, for any fixed $p \in M$,

$$S_p = \{(x,c) \mid x \in M, \ \ c \text{ is a continuous curve from } x \text{ to } p\}.$$

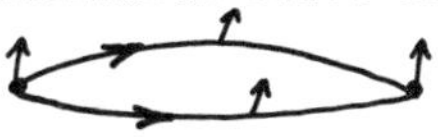

Put:

$$(x,c) \sim (y,c') \iff \begin{cases} x=y \text{ and the time direction is} \\ \text{preserved on the path from } p \text{ to } x \\ \text{to } p \text{ using } c \text{ and } c' \text{ .} \end{cases}$$

We assume of course that g has been used to provide a line element field on M and that a choice of forward direction for time has been made, though in general a continuous such choice may be impossible. The factoring $\tilde{M} = S_p/\sim$ then effectively eliminates these discontinuities. Since M is by hypothesis connected it is arc connected and so the choice of base point p is immaterial, up to diffeomorphism. Evidently, in a physical sense, $\tilde{M}$ represents the same universe as M.

(ii) If (M,g) is time-orientable then $\tilde{M}$ is diffeomorphic to M, because the curves in S_p are factored out by $\sim$. The differentiable structure is provided on $\tilde{M}$ in a similar way to that employed in Theorem 5.

(iii) Observe that g induces a line element field L on M and, by Theorem 6, L lifts to $\tilde{L}$ on $\tilde{M}$; also $\tilde{M}$ is paracompact and hence by Theorem 1 in §1 $\tilde{L}$ induces many Lorentz structures on $\tilde{M}$. However, we can use g directly to determine a unique Lorentz structure $\tilde{g}$ on $\tilde{M}$ because our construction of $\tilde{M}$ uses copies of charts only trivially different from those on M.

(iv) The construction of $(\tilde{M},\tilde{g})$ ensures that it is the smallest time-orientable covering manifold of (M,g), since every other candidate also covers $(\tilde{M},\tilde{g})$.

At each point x in a spacetime (M,g) the Lorentz structure allows a splitting of T_xM into timelike, null and spacelike

directions (cf. §1 above). Hence at each $x \in M$ we can choose a spacelike triad, an ordered triple of linearly independent spacelike vectors, because in the presence of g a timelike distribution determines an orthogonal spacelike one. Next we can enquire if such a choice can be made continuously over M in such a way as to preserve orientation of the spacelike triads round all closed curves, if so then (M,g) is space-orientable. It turns out (cf. Geroch [39]) that time- and space-orientability are independent of one another, and also independent of the physical notion of charge-orientability arising from the transport of elementary charges around closed curves. There are theoretical reasons for believing that physical interactions are charge-parity-time-invariant; this means that either all of charge-, space- and time-orientation reverse round a closed curve or none of them does. The belief is based on the Charge Parity Time or CPT theorem (cf. Geroch [39] p. 84 and references there) and encouraged by experimental evidence. It follows that if we wish to impose time-orientability on (M,g) then to preserve CPT-invariance we must also have space-orientability, and charge orientability; in that case M is orientable.
The mathematical situation is revealed by the following result:

2.8 Theorem

Let M be a connected paracompact n-manifold having a distribution L of dimension p and a distribution S of dimension $(n-p)$ such that TM is spanned by $L \oplus S$. Then any two of the following statements implies the third :

(i) M is orientable ,

(ii) L is orientable ,

(iii) S is orientable .

Proof

Either using algebraic topology or by comparing volume forms for orientable distributions (cf. Theorem 3); both methods are outlined by Whiston [104]. □

Corollary

For a spacetime (M,g) any two of the following statements implies the third:

(i) M is orientable ,

(ii) M is time-orientable ,

(iii) M is space-orientable. □

It is normal to assume that spacetime is time-orientable and then at each point $x \in M$ there is determined a division of all non-spacelike vectors into two classes: forward-pointing and past-pointing (we allow only the zero vector to be in both). Then curves leaving from x with tangent vectors in these classes are called forward-going or past-going, respectively. Hawking and Ellis, [43] chapter 6, provides a thorough study of this causal structure by investigating the following possible global constraints on a time-orientable spacetime (M,g) :

(i) chronology condition : absence of closed timelike curves ;

(ii) causality condition : absence of closed non-spacelike curves ;

(iii) stable causality condition : g lies in an open set in the C^0 open topology on metric tensor fields and no metric in this open set admits closed timelike curves.

Now, in general relativity the Lorentz structure g is determined by the disposition of matter. It would therefore be comforting to think that stability in the sense of constraint (iii) is present in such a way as to allow for possible quantum fluctuations in g , when a suitable theory is agreed for their incorporation. For this reason we note the interesting necessary and sufficient condition established by Hawking and Ellis [43] p. 198-201.

2.9 Theorem

A time-orientable spacetime (M,g) satisfies the stable causality condition if and only if there is a real function f on M whose gradient (i.e. df) is everywhere timelike. □

Remarks

(i) The intuitive view of stable causality is that a slight expansion of the cone of forward going curves does not result in any of them returning to an earlier point.

(ii) The function f can be viewed as giving universal time. Hence, if f exists, the forward cone through $x \in M$ (in the direction of df there) meets a surface of constant f through x only at x. The converse, establishing such an f in the presence of stable causality depends on constructing a measure on M, via a partition of unity, so f is not uniquely fixed.

(iii) We know that since M has the Lorentz structure g there is a timelike distribution L on M (by Theorem 1 in §1), though it is not uniquely determined. Since (M,g) is time-orientable, L is orientable and so determined by a nowhere-zero timelike vector field w on M. Now in the presence of g, w determines a unique nowhere-zero timelike 1-form $g_{\downarrow}(w)$ on M (cf. Dodson and Poston [28] p. 263). From Theorem 9 we can deduce that (M,g) satisfies the stable causality condition if $g_{\downarrow}(w)$ can be chosen to satisfy

$$g_{\downarrow}(w) = df$$

for some real function f on M, that is if $g_{\downarrow}(w)$ is an <u>exact</u> 1-form (cf. IV §2.7). For this to happen we certainly want $g_{\downarrow}(w)$ to be <u>closed</u>, that is

$$dg_{\downarrow}(w) = 0 .$$

It is known (cf. Bishop and Goldberg [7] p. 175) that if $g_{\downarrow}(w)$ is closed then <u>locally</u> it is exact, so a suitable function exists on some subset of the domain for each chart from the atlas for M. In fact this is not saying much because we know that there are always normal coordinate neighbourhoods about each point (cf. [28] p. 356).

(iv) For each $t \in \mathbb{R}$ the surfaces $\overleftarrow{f}(t)$ can be viewed as surfaces of simultaneity in spacetime, but this may not be very useful unless some particular f is distinguished by further physical information. If <u>all</u> of the surfaces $\overleftarrow{f}(t)$, $t \in \mathbb{R}$, are <u>compact</u> then they are all diffeomorphic by following the integral curves of $g^{\uparrow}(df)$; this need not be true if some of them are non-compact.

3. PARALLELIZABILITY

The main result that we shall be discussing here is due to Geroch [37]:

3.1 Theorem

A space- and time-oriented spacetime (M,g) is parallelizable if and only if it admits a spinor structure. □

In fact, this does not extend to orientable compact manifolds carrying a Lorentz structure. Examples of compact non-parallelizable spinor spacetimes have been given by Whiston [105], who also provided the following necessary and sufficient condition from algebraic topology.

3.2 Theorem

A compact spinor spacetime is parallelizable if and only if its Pontryagin number (cf. e.g. Spanier [92]) is trivial. □

We need a definition (cf. Geroch [37] and Whiston [105]):

A spacetime (M,g) is said to admit a spinor structure or is a spinor spacetime if we have the following:

(a) a principal fibre bundle $\tilde{O}M$ over M, with structure group $S\ell(2;\mathbb{C})$;

(b) a two-to-one principal fibre bundle morphism

$$\sigma : \tilde{O}M \longrightarrow O^{+}M ,$$

where $O^{+}M$ is the principal fibre bundle of oriented orthonormal frames with structure group $SO^{+}(1,3)$, the proper Lorentz group. (Cf. Porteous [78] p. 161 ; as a matrix group, $SO^{+}(1,3)$ is that connected subgroup of the Lorentz group $SO(1,3)$ having determinant $+1$.)

We do not propose more than a superficial coverage of the background to Theorem 1 because it, and related results, juxtapose two bodies of theory of very different types that we do not have space to develop here. On the one hand there is the mathematical theory of characteristic classes within algebraic topology, and on the other hand there is the physical motivation for spinors within quantum field theory. A feel for the mathematical aspects of the situation is offered through the following notes and references.

(i) The group $S\ell(2;C)$ consists of unimodular 2×2 complex matrices and it is the universal covering group of the proper Lorentz group. (Cf. Gel'fand, Minlos and Shapiro [35].) Physicists view $S\ell(2;C)$ as the transformation group of spinors, i.e. of certain elements from a 2-dimensional complex vector space (cf. Penrose [75], Pirani [76], Crumeyrolle [19] and Cartan [10].)

(ii) One necessary and sufficient condition for a time- and space-oriented (M,g) to admit a spinor structure is the following isomorphism of fundamental groups:

$$\pi_1(O^+M) \cong \pi_1(M) \oplus \pi_1(SO(1,3)). \quad (\text{N.B. } \pi_1(SO(1,3)) = Z_2)$$

Another such condition is that each of its covering manifolds admits a spinor structure. From the results in §2 we see that, unlike non-orientability, failure to admit a spinor structure cannot be remedied by passing to a covering space. (Cf. Geroch [37].)

(iii) The hard part of Theorem 1 is in constructing a parallelization from a spinor structure. Just why it goes through cannot be better put than in Geroch's own words ([37] p. 1743): "The theorem depends critically on the vanishing of homotopy groups of the spinor group $S\ell(2;C)$. The first homotopy group vanishes essentially because a spinor structure is defined by the property that its fiber is the universal covering space of the fiber of the bundle of frames. (Taking the universal covering group automatically annihilates the first homotopy group.) The second homotopy group vanishes for all the spin groups (in fact, for all Lie groups). The third homotopy group fails to vanish, but at this point we are sufficiently close to the dimension of the manifold that the obstruction to extending a cross section can be made to vanish. Thus the dimension of the manifold enters in an essential way. In fact, the theorem is true in four dimensions, uninteresting in lower dimensions (in this case, every orientable manifold is parallelizable), and false in higher dimensions." (Cf. also Lee [63] p. 425 footnote † and

Clarke [14].)

(iv) In a sequel to his paper quoted above, Geroch [38] found other tests for the admissibility of spinor structures. On the geometrical side, and particularly interesting for the physics of general relativity, is the result that "a certain threshold of curvature must be exceeded before there can be even the possibility of a spacetime having no spinor structure." This is surprising in view of the known topological character of conditions for parallelizability. Also, Geroch gave a topological condition for a time- and space-oriented spacetime (M,g) to be parallelizable. It is sufficient that M is topologically the product of $\mathbb{R}$ with a spacelike 3-surface; hence if M is constructible from initial data on a Cauchy surface (cf. Hawking and Ellis [43] p. 201 et seq.) then it is parallelizable. At work here is the trivial parallelizability of $\mathbb{R}$, and the parallelizability of any orientable 3-manifold (cf. Steenrod [96] p. 203 and 221). In view of Theorem 9 in the previous section we have:

3.3 <u>Theorem</u>

A space- and time-orientable spacetime (M,g) is parallelizable if it satisfies the stable causality condition and if a universal time function $f : M \longrightarrow \mathbb{R}$ so determined has homeomorphims

$$\overset{\leftarrow}{f}(t) \simeq \overset{\leftarrow}{f}(t') \qquad \forall t,t' \in \mathbb{R} . \qquad \square$$

<u>Corollary</u>

The required homeomorphisms are necessarily present (and in fact diffeomorphisms) if $(\forall t \in \mathbb{R})$ $\overset{\leftarrow}{f}(t)$ is compact. $\square$

4. PRODUCT SPACETIMES

We make a few points about a simple class of spacetimes and mention some restrictions on the evolution of the topology of spacelike sections.

The simplest candidate for a spacetime (M,g), to model part or all of the universe, is a product of Riemannian manifolds (S,g_σ), $(\mathbb{R},g_\tau)$ in the form

$$M = \mathbb{R}\times S , \quad g = (-g_\tau) \otimes g_\sigma .$$

Here, S is some 3-manifold representing the space of positions at a given instant of time. This is a simple model because both the topology and the geometry of the space of positions remains unaltered throughout all time, representing a static universe.

The next level of complexity is to keep the product topology, $M = \mathbb{R}\times S$, but to relax the product geometry by taking

$$g \neq (-g_\tau) \otimes g_\sigma .$$

Typically, the geometry of the spacelike sections $\{t\}\times S$ are made to depend on t and the simplest way is to introduce a conformal factor (cf. Hawking and Ellis [43]) to give

$$g = \rho^2.(-g_\tau) \otimes g_\sigma$$

where ρ is some positive function (of time $t\in\mathbb{R}$ only, here) which intuitively serves to measure the 'radius' of space $\{t\}\times S$ at each instant $t\in\mathbb{R}$. Thus, to model the observed Hubble redshift of stellar spectra we represent a universal expansion of space by choosing ρ to be a strictly increasing function of time. Particular examples of spacetimes of these types and others conformal to products are discussed in Hawking and Ellis [43] Ch. 5, (cf. also Misner, Thorne and Wheeler [72], Sachs and Wu [82] and Clarke [17]).

From our previous results we collect the following.

4.1 Theorem

Let a spacetime (M,g) have M diffeomorphic to a product of $\mathbb{R}$ (representing time) with a 3-manifold S (representing space). Then we can deduce:

(i) If S is orientable then S is parallelizable so LS is trivial; also M is orientable, time-orientable and space-orientable and parallelizable hence LM is trivial.

(ii) If S is not orientable then S is not parallelizable and so LS is non-trivial, M is not orientable, not space-orientable, not parallelizable and LM is non-trivial.

(iii) (M,g) satisfies the stable causality condition.

(iv) For each $t \in \mathbb{R}$, $\{t\} \times S$ is a Cauchy surface.

Proof

(i) Since $\mathbb{R}$ is orientable, if S is orientable then M is orientable; and M is time-orientable, therefore by Theorem 8 in the previous section, M is space-orientable. Any orientable 3-manifold is parallelizable (cf. Steenrod [96] p. 203, 221). Finally, LS is trivial if S is parallelizable by Theorem 3 Ch. IV §2.3.

(ii) If S is parallelizable then so is M because $\mathbb{R}$ is parallelizable. Being parallelizable, M is orientable; but since $\mathbb{R}$ is orientable, M is time-orientable and therefore also space-orientable. Again, parallelizability of M ensures triviality of LM.

If S is not orientable then M is neither orientable nor space-orientable. Since any parallelizable manifold is orientable, S and M are not parallelizable and hence LS and LM are non-trivial.

(iii) Since $M = \mathbb{R} \times S$ is time-orientable, stable causality is assured by Theorem 9 in §2 if we can find a universal time function. Such a function is given by projection onto the $\mathbb{R}$ component.

(iv) The subsets $\{t\} \times S$ for $t \in \mathbb{R}$ are the prototypes of all Cauchy surfaces (cf. Hawking and Ellis [43] p. 205). For given $t_o \in \mathbb{R}$ then any $(t',x') \in \mathbb{R} \times S$ lies on the timelike curve

$$c : \mathbb{R} \longrightarrow M : t \longmapsto (t,x')$$

through $(t_o,x') \in \{t_o\} \times S$,

and each inextensible non-spacelike curve through (t',x') meets $\{t\} \times S$ just once. □

Evidently, if $M \equiv \mathbb{R} \times S$ and S is orientable then M has a trivial structure for its bundles because by parallelizability we deduce that

$$LM = \mathbb{R} \times S \times G\ell(4;\mathbb{R}) .$$

Furthermore, by Corollary 1 of Theorem 1 in §1, LM is reducible to the orthonormal frame bundle OM, with structure group the

Lorentz group.

Typical candidates for the Riemannian 3-manifold S in a product spacetime with $M \equiv \mathbb{R}\times S$ are $\mathbb{R}^3$, S^3 and possibly also products like $S^2\times\mathbb{R}$, $S^1\times S^1\times\mathbb{R}$. Through a study of the homotopy groups associated with the classical groups (orthonormal, unitary and symplectic) Husemoller [52] Ch. 7 gives a classification of all vector bundles over spheres of dimension up to 4. Topologically:

$$S^1 \simeq SO(2), \mathbb{R}P^3 \simeq SO(3) \simeq S^3/Z_2$$

$$S^3 \simeq SU(2), SO(4) \simeq S^3\times SO(3)$$

and we have the fundamental groups :

$$\pi_1(SO(2)) = Z , \quad \pi_1(SO(3)) = Z_2 = \pi_1(SO(4)) ,$$

also, of course,

$$\pi_1(S^1\times S^1) = Z\oplus Z , \quad \pi_1(S^3) = \pi_1(\mathbb{R}^n) = \pi_1(S^2) = 0 .$$

Husemoller's classification is through a consideration of bundles having the classical groups as fibres. Whiston [105] has also studied the Lorentz group and the associated spinor groups; here we note his observation that the inclusion

$$SO(3) \hookrightarrow SO^+(1,3)$$

which we encountered earlier is actually a homotopy equivalence and so induces an isomorphism

$$\pi_1(SO^+(1,3)) \equiv \pi_1(SO(3)) = Z_2 .$$

In fact, much material is available generally on the classification of the product spacetimes of likely interest, because of the work that has been done for 3-manifolds. Even with relatively simple topologies there is of course considerable scope left for supplying interesting geometries through the Lorentz structure. This is illustrated by the analyses in Hawking and Ellis [43] of spacetimes having Cauchy surfaces.

It is natural to enquire whether a change in the topology of spacelike regions can occur with time in a physically reasonable, general spacetime. In fact, the question is only meaningful in those spacetimes that admit a splitting into space and time in the locality of interest. One technical construction that is used in these situations is a partial Cauchy surface. We do not propose to become involved with the details, for they are thoroughly covered by Hawking and Ellis [43] and similar methods are used in the references we mention below. We shall merely describe informally the main results that restrict unusual evolution of the topology of spacelike regions. Logically this material fits here because it pertains to spacetimes that are natural generalisations of global products. Additionally, we shall see that under reasonable conditions a change in the topology of a spacelike region will result in some curve becoming inextensible while still in some sense finite; we conclude that it meets a singularity. Singularities are the subject of the next section.

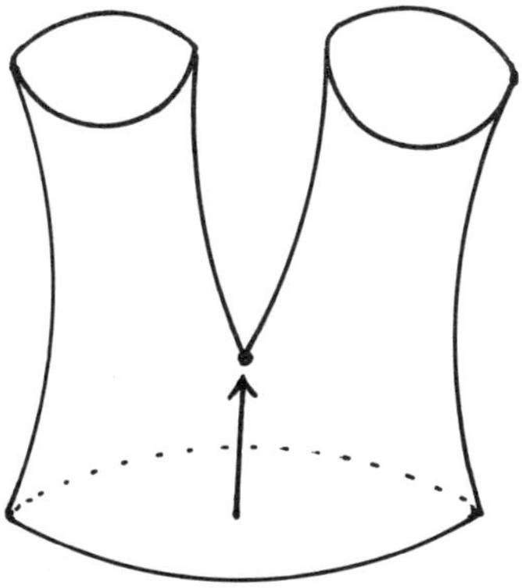

Geroch [36] showed that a change of topology of a compact spacelike region with time in a spacetime is impossible if we do not allow closed timelike curves. Also, Lee [61] established that if a region of space can be enclosed by a spacelike C^2-embedding of S^2 having one congruence (ingoing, say) of null vectors orthogonal to it converging (a uniformly convex sphere), then that region is compact and simply connected; moreover, modulo the Poincaré conjecture, it is homeomorphic to the unit 3-ball. Conversely, if some non-trivial topology of space could be enclosed by a uniformly convex sphere then one of the ingoing null geodesics orthogonal to the enclosing sphere would be expected to be inextensible after a time of the same order as the diameter of this sphere. Physically, an ingoing photon would cease to exist after a finite time. Precisely this happens in the case of the uniformly convex spheres enclosing a Schwarzschild-Kruskal black hole. These spheres have non-compact interior and hence a non-trivial topology. Calculations (cf. Misner, Thorne and Wheeler [72]) show that for a black hole of solar mass (about 10^{30} kg) the radius of the sphere coinciding with the 'event horizon' is a few kilometres, so an ingoing photon passing through the event horizon meets the central singularity after a flight time of about ten microseconds.

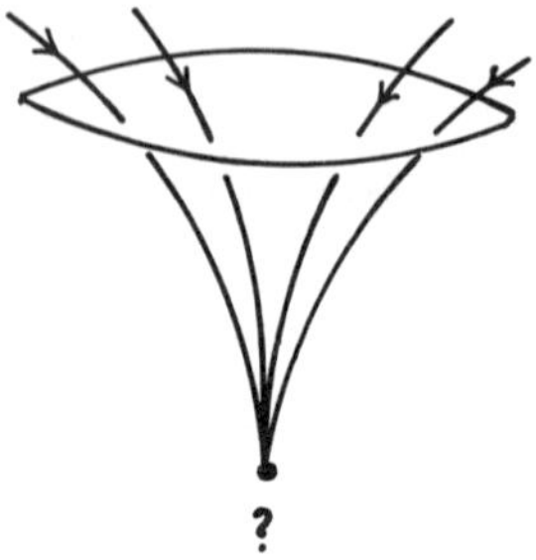

Tipler [98] extended Geroch's result by showing that the causality condition was not necessary if gravity is assumed to be everywhere attractive and subject to Einstein's equation. Specifically, if this latter assumption is true on a compact subset B bounded by the disjoint union of two compact 3-manifolds S_1 and S_2 then $S_1 \equiv S_2$ and $B \simeq S_1 \times [0,1]$. Moreover, if a topology change does occur in a 4-dimensional region B , then B is not compact and it contains a singularity if B is 'finite'. Lee [62] showed that closed and bounded regions of space cannot change topology with time in a spacetime that is timelike and null geodesically complete with gravity everywhere attractive and no closed timelike curves.

Everyday experience suggests to us that in small enough regions we can decompose the observable universe into a succession of spacelike slices parametrized by time, and the spacelike sections have persisting, trivial topologies. More sophisticated experiments interpreted through very reliable theories concerning the behaviour of matter, have convinced cosmologists that the observable universe does indeed contain singularities in the classical sense. We end this section with notes on the three principal cases.

(i) There is overwhelming evidence to support the view that our universe had a 'beginning' some 10^{10} years ago. In particular, certain photons (collected in detectors of the isotropic background black body radiation) yield examples of finite null geodesics that are past-inextensible (beyond the primeval Big Bang).

(ii) The final state of some stars is expected to be a black hole, achieved through a collapse of molecular, atomic and nuclear matter under gravitational forces. Astronomers are convined that no alternative explanation remains for the superdense dark companion of Cygnus Xl.

(iii) The turbulent conditions during the early stages of the universe could have generated black holes over a range of masses. Recently, Hawking has shown that certain quantum theoretical effects are particularly important for small black holes : they 'evaporate' more quickly than large ones. The process is one of losing energy by creating particles in the strong fields surrounding their event horizons. So eventually black holes disappear completely because the rate of energy loss continues to increase, but the process is very slow indeed for large black holes. Now, we believe that the Big Bang was about 10^{10} years ago and this turns out to be the expected liftime of a black hole of mass about 10^{12} kg; it has an event horizon commensurate with nuclear radii. These 'mini black holes' have not yet been detected, but the theoretical basis for their existence continues to be strengthened.

5. SINGULARITIES

We found in the previous section that rather mild topological anomolies could lead to incompleteness and singularities. Hawking and Ellis [43] show quite generally that, in any physically reasonable spacetime, timelike or null geodesic incompleteness is very likely. In relativity theory, timelike and null geodesics are viewed as potential trajectories for 'free' material particles and photons, respectively. By 'free' is meant an absence of external forces. Experimentally, a freely falling satellite seems to follow very precisely a timelike geodesic in a Schwarzschild spacetime used to model a neighbourhood of the Earth. However, a rocket used to place the satellit in orbit

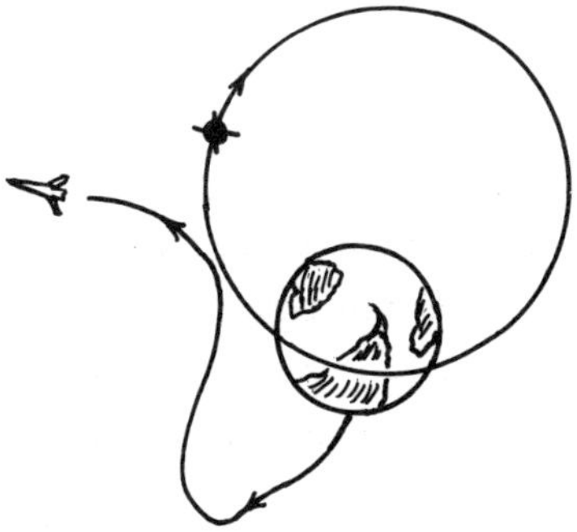

is certainly not falling freely and it follows a curve that is timelike but not necessarily a geodesic. In fact, whereas timelike or null geodesic incompleteness would certainly be sufficient for a spacetime to admit singularities in a physical sense, it may not be necessary. For there is an example of a timelike and null geodesically complete spacetime that contains incomplete timelike curves which could represent the trajectory of a rocket having bounded acceleration. This is in marked contrast to the situation for Riemannian manifolds, where geodesic completeness is actually equivalent to metric completeness (cf. Theorem 4 in Ch. IV §2.8). There is of course an immediate mathematical solution : from Theorem 2 in §1 every spacetime is paracompact and therefore admits a Riemannian structure. The deficiency in this solution is the non-uniqueness of the Riemannian structure, because of the freedom of choice available for the partition of unity on which it is based. Another way to obtain a Riemannian structure h on a spacetime (M,g) with a timelike vector field v is to invert the formula of Avez [2], which we used in §1. If we suppose that

$$g = h - \frac{2}{h(v,v)} h_{\downarrow}(v) \otimes h_{\downarrow}(v) ,$$

then

$$g(a,v) = -h(a,v)$$

and it follows that there is a unique Riemannian structure

$$h = g - \frac{2}{g(v,v)} g_{\downarrow}(v) \otimes g_{\downarrow}(v) .$$

However, this structure depends on the choice of the timelike vector field v and so in a physical sense depends on the existence and distinction of some matter field. Of course many simple spacetimes do have such fields but in general arguments, as far as practicable, we avoid the introduction of extra structure.

Various formulations of criteria for completeness of spacetimes have been tried (cf. Hawking and Ellis [43]) and each furnishes an implicit definition of what is to be a singularity. The strongest competitor on the grounds of mathematical and physical elegance was that of Schmidt [84] (cf. also [41], [81], [86] and [18]). His criterion for curves is called <u>bundle-completeness</u> or <u>b-completeness</u>.

It decides unambiguously which topologically inextensible curves in a spacetime are finite. The test is the length of their horizontal lifts in the frame bundle via the metric $\hat{g}$ induced by the Levi-Cività connection (the metric $\hat{g}$ appeared in the proof of Theorem 5 in Ch. IV §2.8). The great beauty of this concept is that it uses only the connection, which physically is more fundamental even than the Lorentz structure itself because in principle it is directly observable through the behaviour of free particles. Further, a b-incomplete spacetime (M,g) can be provided with a unique b-boundary ∂M, making $M \cup \partial M$ a topological space. The same procedure can be applied to a Riemannian manifold; then the manifold with b-boundary coincides with the usual Cauchy completion, giving more evidence of inherent naturality in the procedure. Further geometric properties of the b-completion are discussed in [26].

The topology on $M \cup \partial M$ makes it possible to formulate questions concerning the geometry of spacetime 'near a singularity', i.e. in every neighbourhood of a point in ∂M. We cannot of course expect a spacetime structure for $M \cup \partial M$ because we suppose that (M,g) itself is inextensible; nor can we expect $M \cup \partial M$ to support a manifold structure at all in general. So it remains to enquire what topological properties of M extend to the completion space. It turns out that $M \cup \partial M$ is connected, locally connected, arcwise connected and second countable but, unlike M, it need not be locally compact nor more than a T_o space. The loss of these properties means that a point in the b-boundary may fail to have any compact neighbourhood and may also be inseparable (by disjoint open sets) from other points in $M \cup \partial M$. In particular then, $M \cup \partial M$ need not be a Hausdorff space and this is a serious disadvantage for physical interpretations of singularities.

Two different ways out of this difficulty have already been investigated. One uses the projective limit of a family of b-completions of co-compact sets in M, the other uses a parallelization to modify the metric in the frame bundle. In both cases the resulting completion is Hausdorff. The price of this improvement is on the one hand the intricacy of any actual calculations using the projective limit or on the other hand a restriction to parallelizable spacetimes. In fact the latter is not a real drawback for most physically interesting spacetimes are parallelizable, the problem lies in choosing among many different parallelizations.

A study of bundle completion, including detailed proofs, has been given in a recent monograph (Dodson [26]) and we shall not repeat that material here. Our treatment is intended to depict something of the universal character of both the underlying construction and its two modifications, through the formulation and proof of three theorems. The first theorem shows that bundle completion does provide a boundary for spacetime; then the corollaries give a breakdown of its properties, progressing from good to bad. Theorem 2 provides the projective limit construction and its consequences, then Theorem 3 deals with the case of parallelizable spacetimes.

This final section of the book is particularly open-ended (cf. Clarke and Schmidt [18] also Tipler, Clarke and Ellis [99]) and no doubt interesting topological and geometrical results will be found along all three lines of attack on spacetime singularities that we describe below. Personally, I should like next to have some classification of parallelization completions and to know when they coincide with other completions. Also, in another direction, I expect some physically interesting completions to arise from 2-jet bundles (cf. Yano and Ishihara [107] Ch. X, also Radivoiovici [80], Hermann [45] and Hennig [44]) because of the prominent place that second order differential equations have in physics.

5.1 Theorem

Given a spacetime (M,g) there is a unique topological space $\bar{M}$, its b-completion, in which M is dense. The set of completion points

$$\partial M = \bar{M}\backslash M$$

its b-boundary, consists of endpoints for inextensible curves in M that have finite horizontal lifts in the frame bundle metric $\hat{g}$ induced by the Levi-Cività connection.

Proof

In the presence of the pseudo-Riemannian structure g we can work in a connected component O^+M of the orthonormal frame bundle, with structure group $SO^+(1,3)$ by Theorem 1 Corollary 1 in §1. Hence, by Theorem 5 in Ch. IV §2.8, we use the closedness of the Lorentz group in $G\ell(4;\mathbb{R})$ to obtain a Riemannian structure on LM and we shall denote its restriction to O^+M by $\hat{g}$. Also, we shall abbreviate $SO^+(1,3)$ to $\hat{O}^+$. The construction then proceeds as follows (cf. [26] for details).

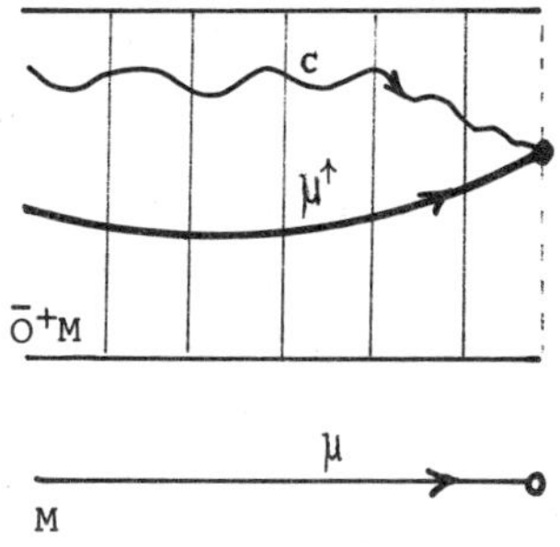

(i) The action of $\hat{O}^+$ on O^+M is uniformly continuous with respect to the topological metric $d_{\hat{g}}$ determined by the Riemannian structure $\hat{g}$. It follows that this action has a unique extension to the Cauchy completion metric space $(\bar{O}^+M, \bar{d}_{\hat{g}})$, and we define $\bar{M}$ to be the quotient $\bar{O}^+M/\hat{O}^+$.

(ii) Points in $\bar{O}^+M\backslash O^+M$ are equivalence classes of Cauchy sequences in O^+M, each of which is representable by a piecewise C^1 curve

$$c : [0,1) \longrightarrow O^+M$$

with bounded length

$$\int_0^1 \hat{g}(\dot{c},\dot{c})^{\frac{1}{2}} < \infty$$

but without a continuous extension in O^+M to domain $[0,1]$.

(iii) For a curve such as c in (ii), its projection onto M, given by

$$\mu : [0,1) \longrightarrow M : t \longmapsto \Pi_{O^+}\circ c(t) ,$$

is likewise inextensible in M to domain $[0,1]$. Further, given any $u \in \overleftarrow{\Pi}_{O^+}\circ\mu(0)$, the (unique) horizontal lift $\mu_u^{\uparrow}$ of μ through u has bounded length in $(O^+M,\hat{g})$. Then $\mu_u^{\uparrow}$ determines a point in the boundary $\bar{O}^+M\backslash O^+M$ and it turns out that this point lies in the same orbit of $\hat{O}^+$ as did the endpoint of c.

(iv) The boundedness of $\overset{\uparrow}{\mu}_u$ does not depend on the choice of $u \in \overset{\leftarrow}{\Pi}_{O^+} \circ \mu(0)$ through which the lift is made.

(v) Given any $z \in \partial M = \bar{M} \backslash M$, every neighbourhood of z in the quotient topology of $\bar{M}$ contains points in M on a curve like μ in (iii). Hence M is dense in $\bar{M}$ and so ∂M is truly a boundary of M . □

<u>Corollary 1</u>

A spacetime (M,g) has $\partial M = \emptyset$ if and only if $(O^+M, d_{\hat{g}})$ is Cauchy complete. □

We say that (M,g) is <u>b-complete</u> if $\partial M = \emptyset$ and <u>b-incomplete</u> if $\partial M \neq \emptyset$. We have earlier remarked that there is a geodesically complete spacetime having incomplete other curves. So geodesic completeness is insufficient to guarantee b-completeness. However, it does work the other way.

<u>Corollary 2</u>

If (M,g) is b-complete then it is also geodesically complete.

<u>Proof</u>

If $\partial M = \emptyset$ then also $(O^+M, \hat{g})$ is complete and so all of its finite curves have endpoints in O^+M . In particular, any horizontal curve of finite length in O^+M has endpoints in O^+M . Now, it is known that every geodesic in (M,g) is the projection of an integral curve of some standard horizontal vector field. Since these integral curves must be complete, their underlying geodesics must be capable of extension to infinite parameter values. □

<u>Corollary 3</u>

Every point x in spacetime has a neighbourhood E_x which is b-complete.

<u>Proof</u>

Choose E_x to be a subset onto which the exponential map at x is a diffeomorphism (cf. [26] p. 356 for details). Schmidt [85] has shown that the b-boundary of E_x coincides with its topological boundary. □

Corollary 4

Plainly the process of b-completion is applicable also to Riemannian manifolds, then the b-completion coincides with the Cauchy completion. (Equally, the process is applicable to any manifold with connection.)

Proof

If (M,g) is a connected Riemannian n-manifold then we again have an orthonormal bundle O^+M, but with structure group $SO(n)$, and on it a Riemannian structure $\hat{g} = \theta\cdot\theta + \omega\cdot\omega$.

For any $x \in M$, $u \in \overleftarrow{\Pi}_{O^+}(x)$ and any

$$X = X_H \oplus X_O \in T_u O^+ M ,$$

we find, by definition of $\hat{g}$ (cf. Theorem 5 in Ch. IV §2.8)

$$\hat{g}(X,X) = \|\theta(X_H)\|^2 + \|\omega(X_O)\|^2 .$$

And since u is orthonormal in g,

$$g(D_u\Pi_{O^+}X, D_u\Pi_{O^+}X) = \|\theta(X_H)\|^2 .$$

So, lengths of curves in (M,g) coincide with their horizontal lifts in $(O^+M,\hat{g})$. Also, any other curves in the bundle O^+M between two given fibres are longer than horizontal curves. Hence the distance between fibres in O^+M coincides with the distance between their underlying points in M. The result follows. □

Corollary 5

$\bar{M}$ need not be locally compact.

Proof

(Cf. Schmidt [84] and Hawking and Ellis [43] p. 283.)

We apply Corollary 4 to the Riemannian submanifold

$$M = \mathbb{R}^2 \setminus \{(x,\sin x^{-1}) \mid x \neq 0\} \cup \{(0,y) \,\big|\, |y| \leq 1\}$$

of two dimensional Euclidean space. In fact, the b-boundary of M^-, the connected component of $(0,-2) \in M$, is

$$\partial M^{-} = \{(x, \sin x^{-1}) \,|\, x \neq 0\} \cup \{(0,-1)\} .$$

Hence, $M^{-} \cup \partial M^{-}$ is not locally compact because $(0,-1)$ has no compact neighbourhood. □

Corollary 6

An attempt to recover omitted points by supplying a b-boundary may yield a different space.

Proof

Consider the Riemannian submanifold of $\mathbb{R}^2$,

$$M = \mathbb{R}^2 \setminus \{(0,y) \,\big|\, |y| \leqslant 1\} .$$

For each $(0,y)$ omitted from $\mathbb{R}^2$ to form M , there are two points in the b-boundary ∂M , except where $y = \pm 1$ where there is only one point. □

The other topological problem with $\bar{M}$ is that it need not be Hausdorff. A necessary and sufficient condition for Hausdorffness is closedness of the graph of $\hat{O}^+$ in $\bar{O}^+M \times \bar{O}^+M$, where $\bar{O}^+M$ is the Cauchy completion of O^+M . The proof is given in [26] p. 433, together with details of the failure even of the T_1 separation property in $\bar{M}$ if $\hat{O}^+$ does not have closed orbits in $\bar{O}^+M$. However, more to the point for cosmology is the failure of the bundle completion of popular spacetimes to be Hausdorff.

Corollary 7

The b-completion $\bar{M}$ of a spacetime (M,g) need not be Hausdorff.

Proof

Two celebrated examples are Friedmann spacetime (with $M = \mathbb{R} \times S^3$ representing a universe expanding from an initial Big Bang singularity) and Schwarzschild-Kruskal spacetime (representing the black hole resulting from stellar collapse). The first was discussed at considerable length in [26] and the second was studied in particular by Johnson [55] (cf. also Clarke [16]). □

A simpler example of a non-Hausdorff b-completion is the case of $M = S^1$ with a constant connection (cf. Dodson and Sulley [29]). There the b-boundary is a single point outside S^1 and the only open neighbourhood of this point in the b-completion $\bar{S}^1$ is the whole of S^1. This example precisely illustrates the following sufficient condition for a bundle completion to be non-Hausdorff. For, every curve that circles S^1 indefinitely is b-incomplete, has every point of S^1 as a limit point and ends on the b-boundary. (Cf. also [26] p. 434 §3.5.)

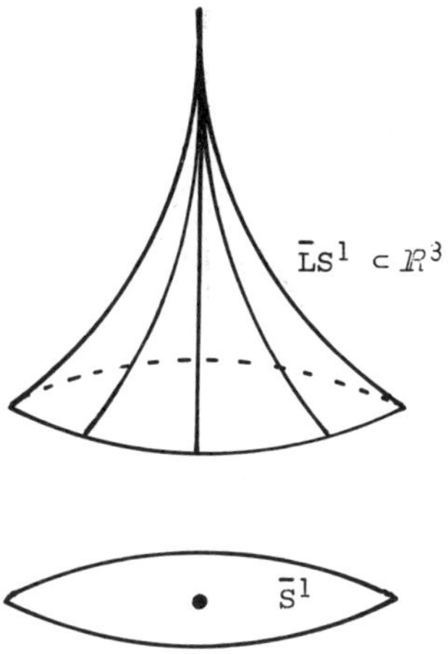

Corollary 8

A point x in a spacetime (M,g) is not Hausdorff separated in $\bar{M}$ from a point $y \in \partial M$ if there is an inextensible curve c in M which has x as a limit point and y as an endpoint.

Proof

(Hawking and Ellis [43] p. 289.)

Given such a curve c then it has a horizontal lift $c^{\uparrow}$ ending at some b in the orbit of $\hat{O}^+$ in $\bar{O}^+M$ over y. If V is an open set in $\bar{M}$ containing y then the bundle of orbits of $\hat{O}^+$ in $\bar{O}^+M$ over V is also open, it contains b and therefore $c^{\uparrow}$ eventually remains in it. Hence V meets every neighbourhood of x because x is a limit point of c. □

We turn now to a modification of the process of b-completion, first suggested by C.J.S. Clarke. It was discussed at the GR8 Conference in 1977 (cf. Clarke [15]) and in a series of subsequent papers (cf. [16], [26], [27], [91] and [99]).

5.2 Theorem

Given a spacetime (M,g) there is a unique topological space $M^{\bullet}$, that is a projective limit formed from b-completions of co-compact subsets of M, in which there is a dense subset M° that is homeomorphic to M. The projective limit boundary

$$\partial^{\bullet}M = M^{\bullet} \backslash M^{\circ}$$

so determined is Hausdorff separated from M° .

<u>Proof</u>

We indicate the steps only, emphasising the categorical aspects of the construction (cf. [27], [91]).

(i) Denote by T the topology on M and also put

$$T^{0} = \{V \in T \mid M\backslash V \text{ is compact}\}$$

$$= \{V_{\alpha} \mid \alpha \in A\} ,$$

for some indexing set A . Then $T^{0} \cup \{\emptyset\}$ is a topology.

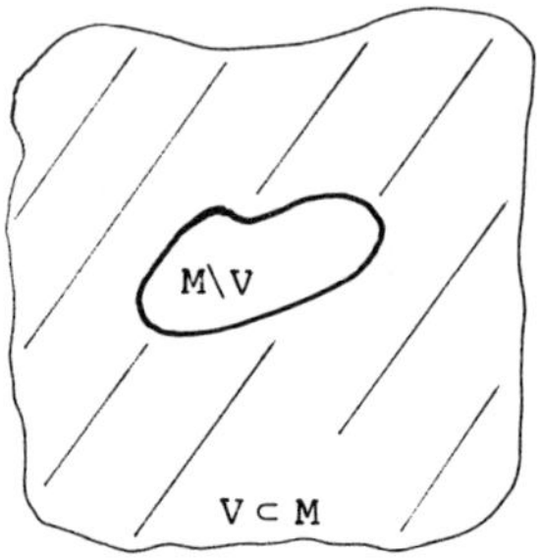

(ii) Denote by $\bar{V}_{\alpha}$ the b-completion of V_{α} and define

$$M_{\alpha} = (M \coprod \bar{V}_{\alpha})/\sim\alpha , \quad \text{for each } \alpha \in A ,$$

to be the pushout (cf. Ch. II §2.3) of the inclusion diagram

$$M \hookleftarrow V_{\alpha} \hookrightarrow \bar{V}_{\alpha} .$$

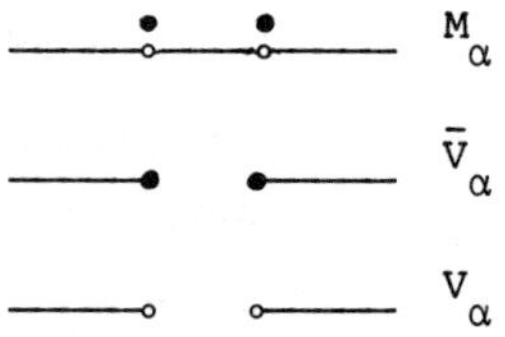

By completeness of Top we know that M_{α} inherits a topology (cf. Ch.II §2.6) unique by being the largest among those that support continuity of projection from $M \coprod \bar{V}_{\alpha}$. Denote this topology by T_{α} .

(iii) The indexing set A has a partial order $\leq$ by

$$\alpha \leq \beta \iff V_{\alpha} \subseteq V_{\beta} .$$

So, for any $\alpha \leq \beta$ we can use the universal property of the pushout squares

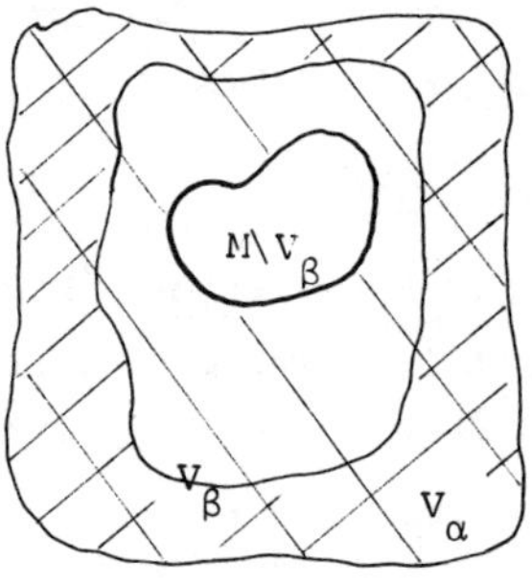

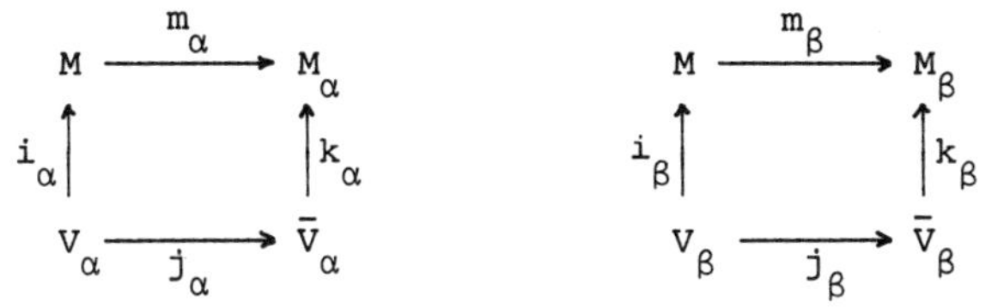

to obtain a unique continuous map $p_{\alpha\beta}$ by

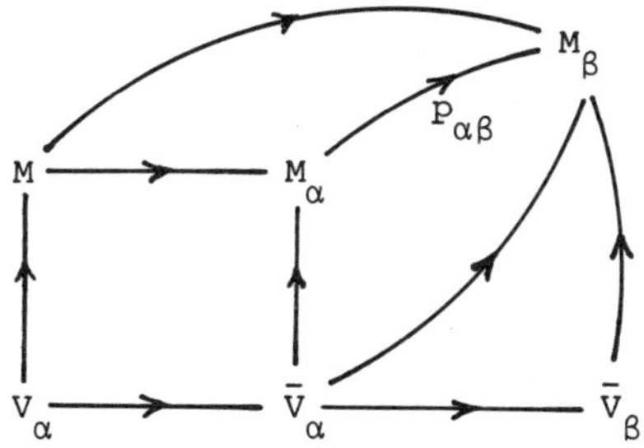

(iv) The diagram so formed in Top :

$$\Delta = \{p_{\alpha\beta} : (M_\alpha, T_\alpha) \longrightarrow (M_\beta, T_\beta) \mid \alpha, \beta \in A \ , \ \alpha \leqslant \beta\}$$

satisfies the requirements in Ch. III §2.8 and so we can form its projective limit space $(M^\bullet, T^\bullet)$, together with a family of continuous projections

$$\chi_\alpha : M^\bullet \longrightarrow M_\alpha : (x_\lambda)_{\lambda\in A} \longmapsto x_\alpha$$

commuting with all

$$M_\lambda \xrightarrow[p_{\lambda\alpha}]{} M_\alpha \xrightarrow[p_{\alpha\beta}]{} M_\beta \ .$$

Note that $M^\bullet$ consists of certain sequences drawn from the M_α , $\alpha\in A$.

(v) Next we use the universal property of the left limit of Δ in (iv). We know (cf. II §2.1) that given any

$$\Delta' = \{f_\alpha : K \longrightarrow M_\alpha \mid \alpha\in A\}$$

in Top, commuting with Δ then there exists a unique continuous map

$$f : K \longrightarrow M^\bullet$$

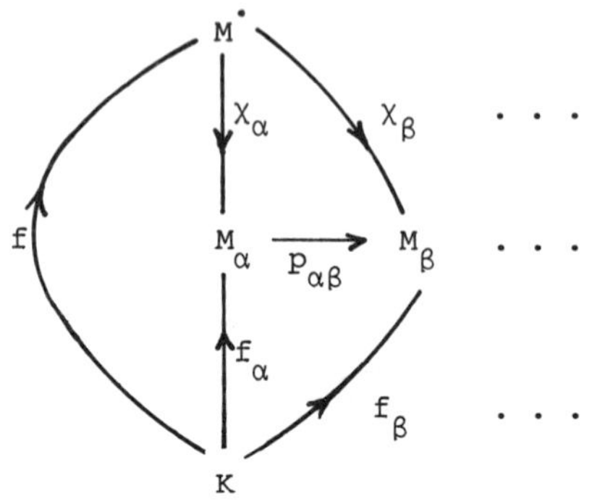

satisfying

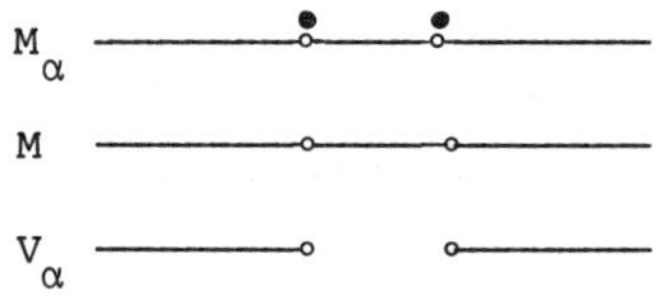

$$f_\alpha = \chi_\alpha \circ f .$$

The first candidate for Δ' is

$$\{M \xrightarrow[f_\alpha]{} M_\alpha \mid \alpha \in A\}$$

where, by our construction, each f_α is the homeomorphism of M onto $f_\alpha M \subseteq M_\alpha$ generated by the injection of M into $M \coprod V_\alpha$. Hence we obtain the unique continuous map

$$f : M \longrightarrow M^{\bullet} : x \longmapsto (x_\alpha = x)_{\alpha \in A}$$

whose image consists of the constant sequences drawn from the M_α , $\alpha \in A$. This f is evidently a bijection and in fact also a homeomorphism onto its image

$$M^{\circ} = fM ,$$

Also, M° turns out to be dense in $M^{\bullet}$. Therefore,

$$\partial^{\bullet} M = M^{\bullet} \setminus M^{\circ}$$

is indeed a fair candidate for a boundary of (M,g) . Quite reasonable separation properties follow (cf. [26] for details):

If $x,y \in M^{\bullet}$ and

<u>either</u> (a) $\chi_\alpha(x)$ and $\chi_\alpha(y)$ are Hausdorff separable in (M_α, T_α) for some $\alpha \in A$,

<u>or</u> (b) $y \in M^{\circ}$,

then x and y are Hausdorff separable in $(M^{\bullet}, T^{\bullet})$. □

So, $(M^{\bullet}, T^{\bullet})$ has the declared properties and we see that intuitively separate points in $\bar{V}_{\alpha}$, the b-completion of some co-compact $V_{\alpha} \in T$, are kept separate in $M^{\bullet}$. In particular, interior points are separated from $\partial^{\bullet} M$. Now, a continuous curve in M that remains in a compact set $M \backslash V_{\alpha}$ may determine a point in the b-boundary ∂M . However, the image in $fM = M^{\circ} \backslash M^{\bullet}$ of such a curve cannot continuously end on the projective limit boundary $\partial^{\bullet} M$. For it too is trapped in a compact set, $f(M \backslash V_{\alpha}) \subset M^{\circ}$, and M° is Hausdorff separated from $\partial^{\bullet} M$.

<u>Corollary 1</u>

(i) Let $c : [0,1) \longrightarrow M$ be a b-incomplete curve not trapped in (or indefinitely often returning to) a compact set, then the curve

$$f \circ c : [0,1) \longrightarrow M^{\circ}$$

has an endpoint in $\partial^{\bullet} M$

(ii) If $x \in \partial^{\bullet} M$ then it is the endpoint of $f \circ c$ for some b-incomplete curve c in M that is not trapped in (or indefinitely often returning to) a compact set.

<u>Proof</u>

The details are rather intricate and can be found in Slupinski and Clarke [91]. We shall just note a further use of the universal property of the projective limit construction (cf. (v) in the proof of the Theorem). In (i) the given curve c eventually remains outside every compact set $M \backslash V_{\alpha}$, so eventually it remains in V_{α} and hence ends on ∂V_{α} . Thus we have continuous

$$c^{\bullet}_{\alpha} : [0,1] \longrightarrow M_{\alpha} , \quad (\forall \alpha \in A) ,$$

commuting with the $p_{\alpha\beta}$ and therefore also a continuous curve

$$c^{\bullet} : [0,1] \longrightarrow M^{\bullet} .$$

But $c^{\bullet}$ coincides with $f \circ c$ on $[0,1)$. □

It is clear that $M^{\bullet}$ is physically more acceptable than $\bar{M}$

as a completion space for spacetimes because the usual topological requirements for distinguishing points persist right up to the boundary in $\overset{\bullet}{M}$. There is another particular advantage, arising from the closed Friedmann spacetimes. These begin with an initial Big Bang singularity, expand, then subsequently collapse into a final singularity after a finite time. In the b-completion the initial and final singularities are topologically identified (full details can be found in [26]). This unphysical behaviour is resolved by the projective limit completion.

Corollary 2

The initial and final singularities in closed Friedmann spacetime are Hausdorff separable in the projective limit completion $\overset{\bullet}{M}$.

Proof

Topologically we have here

$$M \equiv (0,1)\times S^3$$

and geometrically,

$$g = \rho^2.(-g_\tau) \otimes g_\sigma \qquad \text{(cf. §4)}$$

where the conformal factor is typically given by

$$\rho : (0,1) \longrightarrow (0,1) : t \longmapsto 1 - \cos 2\pi t \ .$$

Physically we would expect the initial and final singularities to be separate, but in the b-completion they actually coincide. This does not happen in the projective limit completion. For, removal of a compact set like

$$K_\alpha = [\tfrac{1}{4},\tfrac{1}{2}]\times S^3$$

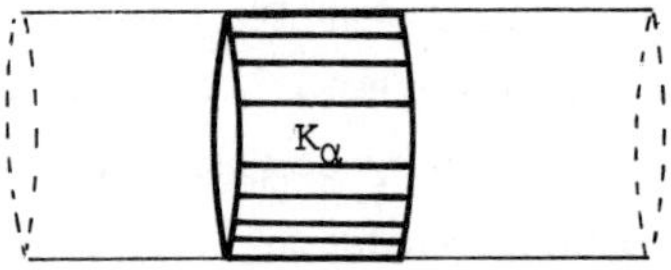

yields a disconnected co-compact set

$$V_\alpha = \left((0,\tfrac{1}{4})\times S^3\right) \cup \left((\tfrac{1}{2},1)\times S^3\right) \ .$$

Then the initial and final singularities are Hausdorff separable in (M_α, T_α) and therefore also in $(\overset{\bullet}{M},\overset{\bullet}{T})$, by the theorem. □

(It is known (cf. [99] §5.2) that the initial and final singularities are not separated in the projective limit completion of the Friedmann spacetime with negative curvature and negative cosmological constant. There, a suitable disconnecting set is not available.)

Finally, we show that similar advantages over the b-completion are available if we are prepared to commit ourselves to a choice of parallelization. We noted earlier (cf. §3) that any physically reasonable spacetime is likely to be parallelizable; the crucial point is an absence of information on how to choose among different available parallelizations. Any information that effectively does distinguish one particular parallelization necessarily imparts extra structure to a spacetime. Physically, extra structure that would fix a choice of parallelization means assuming something like a matter field, which is perfectly reasonable in many situations.

The p-completion available through a parallelization p was discussed at the GR8 Conference in 1977 (cf. [25]) further developed in [26] and shown generally to yield a Hausdorff space by Dodson and Sulley [30]. The motivation for it came from the closed Friedmann b-completion, where initial and final singularities are identified through a sequence of curves that are the result of unboundedly 'high' lifts in the frame bundle. The bundle metric in the p-completion has a vertical term which makes such lifts prohibitively expensive in length and consequently the idenfication is severed. The frame bundle geometry that ensues from using this metric retains many of the good features present in the geometry of Schmidt's metric: The group action remains uniformly continuous; different inner products invoked for the canonical 1-form and connection forms yield a uniformly equivalent metric structure; fibres of the frame bundle are complete, homogeneous spaces. The geometry is studied in [26] , including an analysis of the closed Friedmann spacetime. Here we shall exploit the coincidence, on a parallelization section, of the two alternative Riemannian structures for the frame bundle; further details can be found in [30]. We also show that in at least one case of physical importance the p-completion process gives the usual topological completion when the b-completion process does not.

5.3 <u>Theorem</u>

Given a spacetime (M,g) and a parallelization $p : M \longrightarrow LM$ then there is a unique topological space $\widetilde{M}^p$, its <u>p-completion</u>, in which M is dense. This $\widetilde{M}^p$ is homeomorphic to the Cauchy completion of pM as a Riemannian submanifold of $(LM,\tilde{g}_p)$; in consequence it is a Hausdorff space.

<u>Proof</u>

We know that pM is a diffeomorphic image of M in LM, since we suppose that p is at least C^1. In Theorem 8 of Ch. IV §2.8 we noted the Riemannian structure $\hat{p}$ induced on M by

$$\hat{p} = \hat{g} \circ (D_x p, D_x p)$$

which is, by the Corollary there, also equivalent to

$$\tilde{g}_p \circ (D_x p, D_x p) .$$

Hence we can define $\widetilde{M}^p$ to be the Cauchy completion of the Riemannian manifold $(M,\hat{p})$ viewed as a metric space. Then it necessarily contains M as a dense subset and so we do have

$$\tilde{\partial} M = \widetilde{M}^p \backslash M$$

as a well-defined boundary of M, the <u>p-boundary</u>.

By construction, the diffeomorphism p is also an isometry and so $(M,\hat{p})$ and $pM \subset (LM,\tilde{g}_p)$ are isomorphic in the category of metric spaces, therefore they have homeomorphic Cauchy completions. Like all metric spaces, $\widetilde{M}^p$ has a Hausdorff topology. □

<u>Corollary 1</u>

The p-completion $\widetilde{M}^p$ is homeomorphic to the quotient space $\widetilde{LM}/G$ where $\widetilde{LM}$ is the Cauchy completion of LM in the metric given by the Riemannian structure $\tilde{g}_p$.

<u>Proof</u>

The appropriate curves are constructed in [30]. □

<u>Corollary 2</u>

The topology of the b-boundary of (M,g) may differ from that of its p-boundary.

Proof

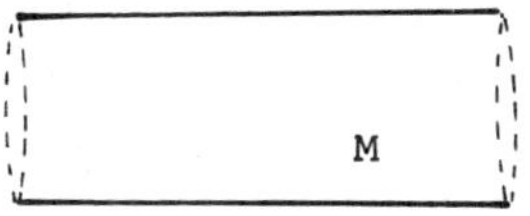

Take $M = (0,1) \times S^1$ and

$$g = (1-\cos 2\pi t)^2 (-g_\tau) \otimes g_\sigma$$

where g_τ and g_σ are the standard Riemannian structures on $(0,1)$ and $S^1 = \mathbb{R}(\text{mod } 1)$, respectively. This is a two-dimensional submanifold of closed Friedmann spacetime, which faithfully carries the crucial geometry - and the 'singularities' at $t = 0,1$.

There is a parallelization, orthonormal in g ,

$$p : M \longrightarrow OM : (t,\sigma) \longmapsto (\partial_t, \partial_\sigma)/(1-\cos 2\pi t) \ .$$

Then it turns out that :

(i) fibres of $\bar{O}M$ over $t = 0,1$ are degenerate (via $\hat{g}$) ;

(ii) fibres over $\tilde{O}M$ over $t = 0,1$ are non-degenerate (via $\tilde{g}_p$) ;

(iii) the singularities at $t = 0,1$ coincide in one point, in the b-completion $\bar{M}$;

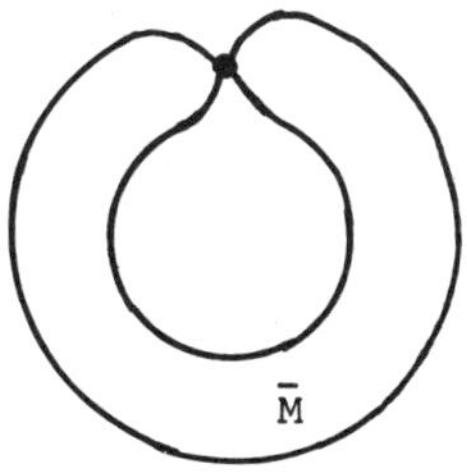

(iv) the singularities at $t = 0,1$ are disjoint copies of S^1 , in the p-completion $\tilde{M}$ and topologically $\tilde{M}$ is the usual completion of $(0,1) \times S^1$, namely

$$\tilde{M} = [0,1] \times S \ .$$

The detailed calculations can be found in [26], based on the original investigations by Bosshard [8] and Johnson [55] of the b-completion of Friedmann spacetime. □

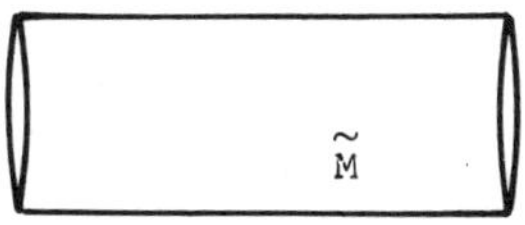

At first sight one might claim that the natural completion for an open cylinder like $(0,1) \times S^1$ is a closed cylinder, as given by $\tilde{M}$ above. However, that is an appeal to naturality

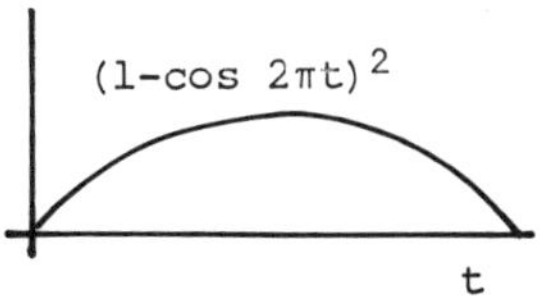

in the category Top - or Met, to be precise. What might not be adequately represented, is the collapse in the pseudo-Riemmanian g of the geometry of space (S^1) as either singularity is approached. The question is really a physical one for the Friedman spacetime : at the instant of creation from the Big Bang, would a particle have a well-defined sense of direction, and in the instant of a final collapse would a particle know from where it had come? In the p-completion $\tilde{M}$, spacelike directions do persist into the singularities; but in the b-completion $\bar{M}$ they do not and that is viewed by Tipler, Clarke and Ellis [99] §5.2, as unreasonable. This defect is shared by the projective limit completion $M^{\bullet}$ - though there at least a cylinder remains a cylinder, instead of becoming a pinched torus. For a particle in our 2-dimensional closed Friedmann spacetime, the available completion spaces may seem like choosing a destiny from among eternal cycling, simple oblivion and organised heaven. If heaven is chosen, then it must be paid for in advance by the selection of a parallelization; random reincarnation or oblivion can be had for nothing.

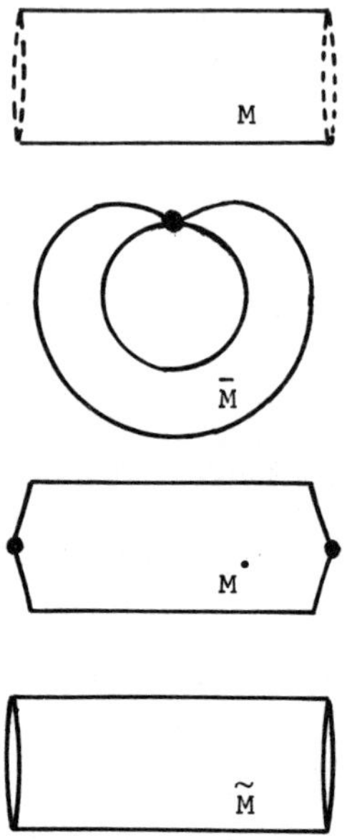

A good place to stop.

Bibliography

1. Arbib, M.A. and Manes, E.G. (1975) Arrows Structures and Functors Academic Press, New York.
2. Avez, A. (1963) "Essais de géometrie Riemannienne hyperbolique globale." Ann.Inst.Fourier 132, 105-190.
3. Bass, R.W. and Witten, L. (1957) "Remark on cosmological models." Rev.Mod.Phys.29, 452-453.
4. Binz, E. and Herrlich, H. (1976) Eds. Categorical Topology Springer-Verlag, Berlin.
5. Birkhoff, G. and Bartree, T.C. (1970) Modern Applied Algebra McGraw-Hill, New York.
6. Bishop, R.L. and Crittenden, R.J. (1964) Geometry of Manifolds Academic Press, New York.
7. Bishop, R.L. and Goldberg, S.I. (1968) Tensor Analysis on Manifolds MacMillan, New York.
8. Bosshard, B. (1976) "On the b-boundary of the closed Friedmann model." Common.Math.Phys. 46, 263-8.
9. Brickell, F. and Clarke, R.S. (1970) Differentiable Manifolds Van Nostrand, London.
10. Cartan, E. (1966) Theory of Spinors MIT Press, Cambridge.
11. Cartan, H. (1971) Differential Forms Kershaw, London.
12. Chevalley, C. (1946) Theory of Lie Groups Princeton University Press, Princeton.
13. Clarke, C.J.S. (1970) "On the global isometric embedding of pseudo-Riemannian manifolds" Proc.Roy.Soc. A314, 417-428.

14. Clarke, C.J.S. (1971) "Magnetic charge, holonomy and characteristic classes," GRG 2, 43-51.

15. Clarke, C.J.S. (1977) "Boundary definitions" Contribution at GR8 Conference; to appear in GRG (1979).

16. Clarke, C.J.S. (1978) "The singular holonomy group" Commun. Math.Phys. 58, 291-297.

17. Clarke, C.J.S. (1979) Elementary General Relativity, Edward Arnold, London.

18. Clarke, C.J.S. and Schmidt, B.G. (1977) "Singularities: the state of the art." GRG 8, 129-137.

19. Crumeyrolle, A. (1969) "Structures spinorielles" Ann.Inst.H. Poincaré A11, 19-55.
(Continued in: "Groupes de spinorialité" Ann.Inst.H.Poincaré A14 (1971) 309-323.)

20. Császár, A. (1978) General Topology Adam Hilger, Bristol.

21. Devlin, K.J. (1977) The Axiom of Constructibility Springer-Verlag, Berlin, Lecture Notes in Mathematics 617.
(Cf. also Fundamentals of Contemporary Set Theory, Springer-Verlag New York 1979)

22. DeWitt, C. and Wheeler, J.A. (1968) Eds: Battelle Rencontres in Mathematics and Physics Benjamin, New York.

23. DeWitt, C. and DeWitt, B. (1964) Eds. Relativity, Groups and Topology Blackie and son, London.

24. Dodson, C.T.J. (1974) "On the Cartan structural equations for the local curvature and torsion of smooth manifolds." Matrix Tensor Q. 25, 1, 20-32.

25. Dodson, C.T.J. (1977) "A new bundle completion for parallelizable spacetimes" Contribution at GR8 Conference; to appear in GRG (1979).

26. Dodson, C.T.J. (1978) "Space-time edge geometry" Int.J.Theor. Phys. 17, 6, 389-504.

27. Dodson, C.T.J. (1979) "Spacetime projective boundaries." Contribution at Bolyai Society Conference; to appear in Proceedings (1980).

28. Dodson, C.T.J. and Poston, T. (1979) Tensor Geometry Pitman, London (2nd edition).

29. Dodson, C.T.J. and Sulley, L.J. (1977) "The b-boundary of S^1 with a constant connection." Lett.Math.Phys. 1, 301-307.

30. Dodson, C.T.J. and Sulley, L.J. (1980) "On bundle completion of parallelizable manifolds." Math.Proc. Camb.Phil.Soc. (in press).

31. Eells, J. (1974) "Fibre bundles" in Global Analysis and its Applications Vol 1 IAEA, Vienna, pp.53-82.

32. Ehresmann, C. (1951) "Les connexions infinitésimales dans une espace fibré differentiable." Colloque de Topologie, Bruxelles, 5-8 June 1950, CBRM (Georges Thone, Liege) pp. 29-55.

33. Flanders, H. (1963) Differential Forms Academic Press, New York.

34. Friedrich, H. (1974) "Construction and properties of space-time b-boundaries." GRG 5, 681-697.

35. Gel'fand, I.M., Minlos, R.A. and Shapiro, Z.Ya. (1963) Representations of the Rotation and Lorentz Groups and their Applications Pergamon, Oxford.

36. Geroch, R.P. (1967) "Topology in general relativity." J.Math. Phys. 8,4, 782-786.

37. Geroch, R.P. (1968) "Spinor structures of space-times in general relativity I." J.Math.Phys. 9,11, 1739-1744.

38. Geroch, R.P. (1970) "Spinor structures of space-times in general relativity II." J.Math.Phys. 11,1, 343-348.

39. Geroch, R.P. (1971) "Space-time structure from a global viewpoint." In Sachs, R.K. Ed. General Relativity and Cosmology (Proc. 1969 Int. School Enrico Fermi) Academic Press, New York, pp. 77-103.

40. Greub, W., Halperin, S. and Vanstone, R. (1972) Connections, Curvature and Cohomology I Academic Press, New York.

41. Hájiček, P. and Schmidt, B.G. (1971) "The b-boundary of tensor bundles over a space-time" Commun. Math.Phys. 23, 285-295.

42. Hano, J. and Ozeki, H. (1956) "On the holonomy groups of linear connections." Nagoya Math.J. 10, 97-100.

43. Hawking, S.W. and Ellis, G.F.R. (1973) The Large Scale Structure of Space-time C.U.P., Cambridge.

44. Hennig, J.B. (1978) "G-structures and space-time geometry I, geometric objects of higher order." Preprint IC/78/46 International Centre for Theoretical Physics, Trieste. (Part II in preparation.)

45. Hermann, R. (1970) Vector Bundles in Mathematical Physics I Benjamin, New York.

46. Herrlich, H. and Strecker, G.E. (1973) Category Theory Allyn and Bacon, Boston.

47. Higgins, P.J. (1971) Categories and Groupoids Van Nostrand Reinhold, London; Mathematical Studies 32.

48. Higgins, P.J. (1974) Introduction to Topological Groups C.U.P., Cambridge; L.M.S. Lecture Notes 15.

49. Hirsch, M.W. (1976) Differential Topology Springer-Verlag, New York.

50. Hochschild, G. (1965) The Structure of Lie Groups Holden-Day, San Francisco.

51. Hodge, W.V.D. (1952) Theory and Application of Harmonic Integrals, C.U.P., London.

52. Husemoller, D. (1975) Fibre Bundles Springer-Verlag, New York (2nd Edition).

53. Ihrig, E. (1976) "The holonomy group in general relativity and the determination of g_{ij} from T^i_j." GRG, 7, 313-323.

54. Jameson, G.J.O. (1974) Topology and Normed Spaces Chapman and Hall, London.

55. Johnson, R.A. (1977) "The bundle boundary in some special cases." J.Math.Phys. 18, 898-902.

56. Kobayashi, S. (1972) Transformation Groups of Differential Geometry Springer-Verlag, Berlin.

57. Kobayashi, S. and Nomizu, K. (1963) Foundations of Differential Geometry I Interscience, New York.

58. Kobayashi, S. and Nomizu, K. (1969) Foundations of Differential Geometry II Interscience, New York.

59. Kowalsky, H-J. (1965) Topological Spaces Academic Press, New York.

60. Lang, S. (1972) Differential Manifolds Addison Wesley, Reading, Massachussets.

61. Lee, C.W. (1976) "A restriction on the topology of Cauchy surfaces in general relativity." Commun. Math.Phys. 51, 157-162.

62. Lee, C.W. (1978) "Topology change in general relativity." Proc.Roy.Soc. A364, 295-308.

63. Lee, K.K. (1973) "Global spinor fields in space-time." GRG 4, 421-433.

64. Lee, K.K. (1975) "On the fundamental groups of space-times in general relativity." GRG 6, 239-242.

65. MacDonald, I.D. (1968) The Theory of Groups Clarendon Press, Oxford.

66. MacLane, S. (1968) Geometrical Mechanics, Mathematics Department Lecture Notes, Univeristy of Chicago.

67. MacLane, S. (1971) Categories for the Working Mathematician Springer-Verlag, New York.

68. Maddox, I.J. (1970) Elements of Functional Analysis C.U.P., Cambridge.

69. Marathe, K.B. (1972) "A condition for paracompactness of a manifold." J.Diff.Geom. 7, 571-573.

70. Markus, L. (1955) "Line element fields and Lorentz structures on differentiable manifolds." Annals.Math. 62, 3, 411-417.

71. Misner, C.W. (1963) "The flatter regions of Newman, Unti and Tambourino's generalized Schwarzschild space." J.Math.Phys. 4, 924-937.

72. Misner, C.W., Thorne, K. and Wheeler, J.A. (1969) Gravitation Freeman, San Francisco.

73. Munkres, J.R. (1954) Elementary Differential Topology Princeton University Press, Princeton.

74. Palais, R.S. (1968) Foundations of Global Non-Linear Analysis Benjamin, New York.

75. Penrose, R. (1968) "The structure of space-time." In C.M. DeWitt and J.A. Wheeler, Eds. Battelle Rencontres in Mathematics and Physics Benjamin, New York; pp. 121-235.

76. Pirani, F.A.E. (1965) in Lectures in General Relativity Brandeis Summer Institute in Theoretical Physics, 1964. Prentice Hall, Eaglewood Cliffs, N.J.

77. Pontryagin, L.S. (1966) Topological Groups Gordon and Breach, New York (2nd edition).

78. Porteous, J.R. (1969) Topological Geometry Van Nostrand Reinhold, London.

79. Poston, T. and Stewart, I. (1978) Catastrophe Theory and its Applications Pitman, London (2nd edition).

80. Radivoiovici, M. (1979) "On the geometry of the tangent bundle of order 2." Contribution at Bolyai Society Conference; to appear in Proceedings (1980).

81. Sachs, R.K. (1973) "Spacetime b-boundaries." Commun.Math. Phys. 33, 215-220

82. Sachs, R.K. and Wu, H. (1976) General Relativity for Mathematicians Springer-Verlag, New York.

83. Schmidt, B.G. (1968) Riemannsche Räume mit mehrfach transitiver Isometriegruppe Doktorarbeit, University of Hamburg.

84. Schmidt, B.G. (1971) "A new definition of singular points in general relativity." GRG 1, 269-280.

85. Schmidt, B.G. (1973) "Local completeness of the b-boundary." Commun.Math.Phys. 29, 49-54.

86. Schmidt, B.G. (1974) "A new definition of conformal and projective infinity of spacetimes." Commun. Math.Phys. 36, 73-90.

87. Schwartz, J.T. (1968) Differential Geometry and Topology Gordon and Breach, New York.

88. Serre, J-P. (1965) Lie Algebras and Lie Groups Benjamin, New York.

89. Simms, D.J. and Woodhouse, N.M.J. (1976) Geometric Quantization Springer-Verlag, Berlin; Lecture Notes in Physics 53.

90. Singer, I.M. and Thorpe, J.A. (1967) Lecture Notes on Elementary Topology and Geometry, Scott, Foresman, Illinois.

91. Slupinski, M.J. and Clarke, C.J.S. (1980) "Projective limits in relativity." Commun.Math.Phys. (in press).

92. Spanier, E. (1966) Algebraic Topology McGraw-Hill, New York.

93. Spivak, M. (1965) Calculus on Manifolds Benjamin, New York.

94. Spivak, M. (1975) Differential Geometry Vol. I Publish or Perish Inc. Brandeis University.

95. Steen, L.A. and Seebach Jr., J.A. (1978) Counterexamples in Topology Springer-Verlag, New York (2nd edition)

96. Steenrod, N. (1955) Topology of Fibre Bundles Princeton University Press, Princeton.

97. Strooker, J.R. (1978) An Introduction to Categories, Homological Algebra and Sheaf Cohomology. C.U.P., Cambridge.

98. Tipler, F.J. (1977) "Singularities and causality violation." Annals.Phys. 108, 1-36.

99. Tipler, F.J., Clarke, C.J.S. and Ellis, G.F.R. (1978) "Singularities and horizons" Preprint University of California to appear in the GRG Einstein Centenary Volume.

100. Wall, C.T.C. (1972) A Geometric Introduction to Topology Addison Wesley, Reading, Massachussetts.

101. Weyl, H. (1922) Space Time Matter Dover, New York.

102. Whiston, G.S. (1974) "Hyperspace cobordism theory of space-time." Int.J.Theor.Phys. 11, 285-288.

103. Whiston, G.S. (1974) "Topics on space-time topology II." Int.J.Theor.Phys. 11, 341-351.

104. Whiston, G.S. (1974) "Topics on space-time geometry." GRG, 5, 525-537.

105. Whiston, G.S. (1975) "Topics on space-time topology III" Int.J.Theor.Phys. 12, 225-240.

106. Whiston, G.S. (1978) "Compact spinor space-times." J.Phys. A. 11, 1203-1209.

107. Yano, K. and Ishihara, S. (1973) Tangent and Cotangent Bundles Dekker, New York.

Index